McGraw-Hill Education

Algebra I
Review and Workbook

Sandra Luna McCune, PhD

New York Chicago San Francisco Athens London Madrid
Mexico City Milan New Delhi Singapore Sydney Toronto

9 LON 23 22

ISBN 978-1-260-12894-9
MHID 1-260-12894-6

e-ISBN 978-1-260-12895-6
e-MHID 1-260-12895-4

Contents

CHAPTER 1

Understanding the Real Numbers

Classifying Real Numbers

The **real numbers** consist of the rational numbers and the irrational numbers. A **rational number** is a number that can be expressed as a quotient of an integer divided by a nonzero integer. (The **integers** are the numbers ..., −3, −2, −1, 0, 1, 2, 3, ...) The decimal representations of rational numbers terminate or repeat. The **irrational numbers** are numbers that cannot be expressed as the quotient of two integers. Their decimal representations neither terminate nor repeat. Frequently, real numbers are called **signed numbers** because they are positive, negative, or zero.

EXAMPLE

- $5.12 \rightarrow$ Rational number
- $-\frac{2}{5} \rightarrow$ Rational number
- $1\frac{1}{2} \rightarrow$ Rational number
- $0.333\ldots \rightarrow$ Rational number
- $0.\overline{14} \rightarrow$ Rational number
- $6.020020002\ldots \rightarrow$ Irrational number
- $\pi \rightarrow$ Irrational number

The three dots indicate that the pattern continues in the same manner without end.

The bar above "14" means that a block of digits continues to repeat without end.

The number π is the ratio of the circumference of a circle to its diameter.

EXERCISE 1.1

Identify the real number as either rational or irrational.

1. $\frac{3}{4}$

2. 0

3. $0.\overline{3}$

4. -9.22

5. $-\frac{2}{5}$

6. $0.454554555...$

7. $-\frac{107}{3}$

8. $0.111...$

9. $100.121212...$

10. $-9.\overline{23}$

11. $-\pi$

12. 0.025

13. $-0.333...$

14. $101{,}001{,}000$

15. $0.010010001...$

16. $\frac{5}{3}$

17. $-\frac{1}{3}$

18. $1\frac{2}{3}$

19. $0.454545...$

20. $-\frac{9}{4}$

Rational and Irrational Roots

A letter that is used to represent a number is a **variable.**

If n is a positive integer, an ***n*th root** of x is a number that when used as a factor n times gives x, the number's ***n*th power**. The notation $\sqrt[n]{x}$ is a **radical**. It indicates an ***n*th root** of the number x. The number n is the **index**, and x is the **radicand**. If n is a positive *even* integer and x is positive, then $\sqrt[n]{x}$ indicates the positive real nth root of x. (Even roots of negative numbers are not real numbers.) If n is a positive *odd* integer and x is any real number, then $\sqrt[n]{x}$ indicates the real nth root of x. A number that is an exact nth power of another number is a **perfect *n*th power**. Roots of perfect nth powers are rational numbers, while roots that cannot be determined exactly are irrational numbers. Decimal representations of irrational numbers are given as approximations. For example, $\sqrt{2} \approx 1.414$.

EXAMPLE

- $\sqrt{3} \rightarrow$ Irrational because 3 is not a perfect square.
- $\sqrt[3]{-8} \rightarrow$ Rational because $\sqrt[3]{-8} = -2$.
- $-\sqrt{36} \rightarrow$ Rational because $-\sqrt{36} = -6$.
- $\sqrt[4]{20} \rightarrow$ Irrational because 20 is not a perfect 4th power.
- $\sqrt[4]{-16} \rightarrow$ Not a real number.

EXERCISE 1.2

Identify the root as either rational, irrational, or not real. Justify your answer.

1. $\sqrt{5}$
2. $\sqrt[3]{8}$
3. $\sqrt{4}$
4. $-\sqrt{4}$
5. $\sqrt{-4}$
6. $\sqrt{16}$
7. $\sqrt[3]{125}$
8. $\sqrt[3]{-125}$
9. $\sqrt[4]{12}$
10. $\sqrt[5]{32}$
11. $\sqrt[5]{-32}$
12. $\sqrt{0.25}$
13. $\sqrt{\frac{9}{4}}$
14. $\sqrt[3]{64}$
15. $\sqrt{100}$
16. $\sqrt{\frac{5}{3}}$
17. $\sqrt[3]{-\frac{1}{3}}$
18. $-\sqrt{625}$
19. $\sqrt[6]{-64}$
20. $\sqrt[3]{75}$

CHAPTER 2

Properties of the Real Numbers

Commutative and Associative Properties

Commutative Property of Addition When you add two numbers, you can reverse the order of the numbers without changing the sum.

Commutative Property of Multiplication When you multiply two numbers, you can reverse the order of the numbers without changing the product.

EXAMPLE

- $6 + 4 = 4 + 6$
- $5.8 + 2.1 = 2.1 + 5.8$
- $3 \cdot 8 = 8 \cdot 3$
- $\left(\frac{1}{2}\right)(10) = (10)\left(\frac{1}{2}\right)$

Generally, in algebra, do not use the times symbol (×) to indicate multiplication. Instead, for numerical quantities, use parentheses or the raised dot multiplication symbol (·).

Associative Property of Addition When you have three numbers to add together, the final sum will be the same regardless of the way you group the numbers (two at a time) to perform the addition.

Associative Property of Multiplication When you have three numbers to multiply together, the final product will be the same regardless of the way you group the numbers (two at a time) to perform the multiplication.

EXAMPLE

$(4.9 + 0.1) + 3.5 = 4.9 + (0.1 + 3.5)$

$(3 \cdot 10)\frac{1}{2} = 3\left(10 \cdot \frac{1}{2}\right)$

EXERCISE 2.1

For 1 to 10, identify which property is represented in the statement.

1. $(5 \cdot 100)\frac{1}{5} = 5\left(100 \cdot \frac{1}{5}\right)$
2. $16 + 24 = 24 + 16$
3. $4 \cdot 75 = 75 \cdot 4$
4. $(1.25 + 0.75) + 6.0 = 1.25 + (0.75 + 6.0)$
5. $\left(\frac{3}{4}\right)(36) = (36)\left(\frac{3}{4}\right)$
6. $\sqrt{5} + 3 = 3 + \sqrt{5}$
7. $(6 + 4) + 20 = (4 + 6) + 20$
8. $(5 + 3) + 2 + 8 = 5 + (3 + 2) + 8$
9. $4 \cdot 3 + 18 = 3 \cdot 4 + 18$
10. $5 \cdot \sqrt{3} = \sqrt{3} \cdot 5$

For 11 to 20, complete the statement using the indicated property.

11. $2 \cdot \frac{3}{4} =$ __________, commutative property of multiplication
12. $(7 \cdot 8) \cdot 5 =$ __________, associative property of multiplication
13. $\sqrt{19} + \sqrt{3} =$ __________, commutative property of addition
14. $4 + 2 + 16 + 8 =$ __________ $+ 16 + 8$, commutative property of addition
15. $(6 \cdot 15) \cdot \frac{1}{3} =$ __________, associative property of multiplication
16. $(1.5)(2) + 2.8 + 0.2 = (1.5)(2) +$ __________, commutative property of addition
17. $(1 + 18) + 2 =$ __________, associative property of addition
18. $(2.25)(-10) =$ __________, commutative property of multiplication
19. $(44 \cdot 3) \cdot \frac{1}{3} =$ __________, associative property of multiplication
20. $(24 + 6)(30) = ($__________$)(30)$, commutative property of addition

Identity and Inverse Properties

Additive Identity Property You have a real number, namely 0, whose sum with any real number is the number itself.

Multiplicative Identity Property You have a real number, namely 1, whose product with any real number is the number itself.

EXAMPLE

- $-25 + 0 = 0 + -25 = -25$
- $\frac{1}{2} + 0 = 0 + \frac{1}{2} = \frac{1}{2}$
- $5 \cdot 1 = 1 \cdot 5 = 5$
- $-\sqrt{2} \cdot 1 = 1 \cdot -\sqrt{2} = -\sqrt{2}$

Additive Inverse Property Every real number has an additive inverse (its opposite) that is a real number whose sum with the number is 0.

Multiplicative Inverse Property Every real number, *except zero*, has a multiplicative inverse (its reciprocal) whose product with the number is 1.

EXAMPLE

- $49 + -49 = -49 + 49 = 0$
- $-\frac{2}{3} + \frac{2}{3} = \frac{2}{3} + -\frac{2}{3} = 0$
- $7 \cdot \frac{1}{7} = \frac{1}{7} \cdot 7 = 1$
- $(-16)\left(-\frac{1}{16}\right) = \left(-\frac{1}{16}\right)(-16) = 1$

EXERCISE 2.2

For 1 to 10, identify which property is represented in the statement.

1. $-28 + 28 = 0$

2. $16 + -16 = 0$

3. $(0.25)(4) = 1$

4. $x \cdot 1 = x$

5. $(1)\left(\frac{2}{3}\right) = \frac{2}{3}$

6. $2 \cdot \frac{1}{2} = 1$

7. $0 + 20 + 8 = 20 + 8$

8. $\left(-\frac{3}{4}\right)\left(-\frac{4}{3}\right) = 1$

9. $4 \cdot 1 + 18 = 4 + 18$

10. $-\sqrt{3} + \sqrt{3} = 0$

For 11 to 20, complete the statement using the indicated property.

11. $\frac{4}{5} \cdot \frac{5}{4} =$ ________, multiplicative inverse property

12. $(7 + 0) \cdot 5 = ($________$) \cdot 5$, additive identity property

13. $\sqrt{19} + -\sqrt{19} =$ ________, additive inverse property

14. $x + 0 + 2.6 + 1.4 =$ ________ $+ 2.6 + 1.4$, additive identity property

15. $\left(7 \cdot \frac{1}{7}\right)(100) = ($________$)(100)$, multiplicative inverse property

16. $-\frac{3}{5} +$ ________ $= 0$, additive inverse property

17. $\left(\frac{9}{10}\right)($________$) = 1$, multiplicative inverse property

18. $\left(-\frac{9}{10}\right)($________$) = 1$, multiplicative inverse property

19. $(1 \cdot 3)\left(\frac{1}{3}\right) = ($________$)\left(\frac{1}{3}\right)$, multiplicative identity property

20. $($________ $+ 24)(30) = (0)(30)$, additive inverse property

Distributive Property

Distributive Property When you have a number times a sum (or a sum times a number), you can multiply each number separately first and then add the products.

A number on the side of parentheses means that the quantity inside the parentheses is to be multiplied by the number.

EXAMPLE

- $3(10 + 5) = 3 \cdot 10 + 3 \cdot 5 = 30 + 15 = 45$
- $(30 + 14)5 = 30 \cdot 5 + 14 \cdot 5 = 150 + 70 = 220$

EXERCISE 2.3

For 1 to 10, use the distributive property to complete the statement.

1. $2(8 + 10) = 2 \cdot 8 +$ ________

2. $4($________$) = 4 \cdot 7 + 4 \cdot 3$

3. $(0.25)(1 + 4) = (0.25)(1) +$ ________

4. $(5 + 8)20 = 5 \cdot 20 +$ ________

5. $(15)\left(\frac{2}{3} +\right.$ ________$\left.\right) = 15 \cdot \frac{2}{3} + 15 \cdot \frac{1}{3}$

6. $2(x + 5) = 2 \cdot x +$ ________

7. $3 \cdot a + 3 \cdot b = 3($________$)$

8. $a($________$) = a \cdot b + a \cdot c$

9. $-4 \cdot 9 + -4 \cdot 11 = -4($________$)$

10. $7 \cdot \frac{1}{2} + 3 \cdot \frac{1}{2} = ($________$)\frac{1}{2}$

For 11 to 20, use the distributive property to evaluate the expression.

11. $2(8+10) =$ ________

12. $4(7+3) =$ ________

13. $(0.25)(1+4) =$ ________

14. $(5+8)20 =$ ________

15. $(15)\left(\frac{2}{3}+\frac{1}{3}\right) =$ ________

16. $0.2(10+5) =$ ________

17. $\frac{3}{4}\left(\frac{4}{3}+\frac{8}{9}\right) =$ ________

18. $8(10+5) =$ ________

19. $(30+2)8 =$ ________

20. $(7+3)\frac{1}{2} =$ ________

Zero Factor Property

Zero Factor Property: If a real number is multiplied by 0, the product is 0; and if the product of two numbers is 0, then at least one of the numbers is 0.

EXAMPLE

- $-18 \cdot 0 = 0$
- $0 \cdot \frac{3}{4} = 0$
- $(7.13)(0) = 0$
- $(9)(1000)(562)(0)(31) = 0$
- If $xy = 0$, then either $x = 0$ or $y = 0$.
- If $3y = 0$, then $y = 0$. (Because, clearly, $3 \neq 0$.)

EXERCISE 2.4

Use the zero factor property to complete the statement.

1. $0 \cdot \frac{7}{8} =$ ________

2. $400($________$) = 0$

3. $(x)(0) =$ ________

4. $0(5+x) =$ ________

5. $(15)($________$)(100)(65) = 0$

6. $(a+b)0 =$ ________

7. $(0.85)(10.25)(3.24)(0) =$ ________

8. $(4.5+9.9-7.5)($________$) = 0$

9. $(0)(-4 \cdot 9 + 3.5 + 1.2) =$ ________

10. $7 \cdot \frac{1}{2} \cdot x \cdot \frac{2}{7} \cdot 0 \cdot \frac{1}{2} =$ ________

CHAPTER 3

The Number Line and Comparing Numbers

The Number Line

The real numbers make up the real number line (or simply, the number line). Every real number corresponds to a point on the number line, and every point on the number line corresponds to a real number.

EXAMPLE

The numbers $-\pi, -1.\overline{3}, -0.5, \frac{3}{5}$, 1.4, and $\sqrt{10}$ are **graphed** on the number line below.

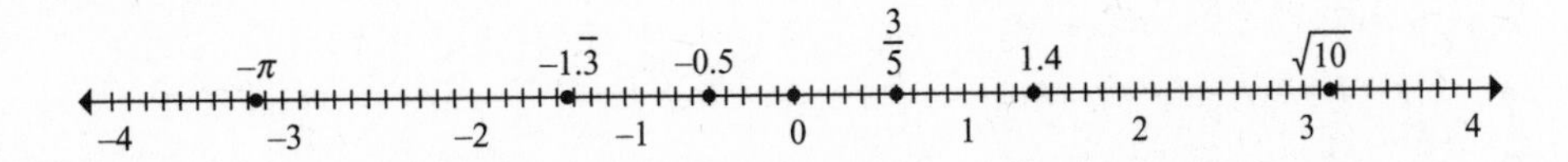

EXERCISE 3.1

Sketch a number line, then graph the points corresponding to the list of numbers given.

1. $-5, -1, 0, 4, 6$
2. $-6, -1, -\frac{1}{2}, 0, 0.\overline{3}$
3. $-4.5, -3, -1\frac{1}{2}, 0.75, 3\frac{1}{4}$
4. $0, 0.5, 2.5, 3.5, 4$

Comparing Numbers

Symbols used in comparing numbers are the following: = (is equal to), ≠ (is not equal to), < (is less than), > (is greater than), ≤ (is less than or equal to), and ≥ (is greater than or equal to). If two numbers coincide, they are equal; otherwise, they are unequal. When you compare two distinct numbers, the number that is farther to the right on the number line is the greater number.

> Two numbers are *distinct* if they are not the same number.

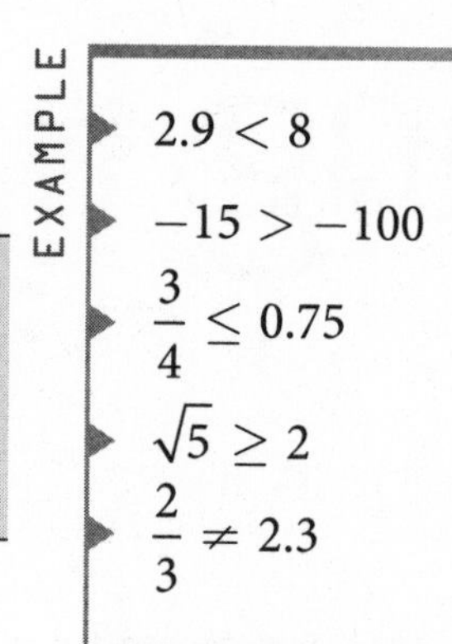

EXAMPLE

- $2.9 < 8$
- $-15 > -100$
- $\frac{3}{4} \leq 0.75$
- $\sqrt{5} \geq 2$
- $\frac{2}{3} \neq 2.3$

> The fraction 3/4 is less than or equal to 0.75 because $\frac{3}{4} = 0.75$

EXERCISE 3.2

For 1 to 15, use a number line to determine whether the statement is true or false.

1. $-100 < -1000$

2. $\frac{1}{2} > 0.2$

3. $0 > 8$

4. $-6 < 10$

5. $3 \leq -3$

6. $-4 \geq -7$

7. $\frac{1}{2} \leq 0.5$

8. $-0.8 < -0.85$

9. $-\frac{4}{5} = 0.8$

10. $9 \leq -9$

11. $-11 < -12$

12. $-\frac{7}{8} > -\frac{3}{4}$

13. $0.25 > 0.3$

14. $0.2 \geq \frac{1}{5}$

15. $0.75 < -1$

For 16 to 20, use a number line to write the list of numbers in order from least to greatest.

16. $0.2, -0.6, \frac{1}{2}, -\frac{2}{3}, \frac{5}{8}$

17. $-3, -8, 0, -2, -2\frac{1}{2}$

18. $-5, 5, -3, -\frac{2}{3}, 8$

19. $0.4, -0.39, \frac{3}{7}, -\frac{1}{3}, 1$

20. $-100, -200, 0, -25, 30$

CHAPTER 4

Absolute Value

Determining Absolute Value

The **absolute value** of a real number is its distance from zero on the number line. The absolute value of a positive number is the same as the number itself. The absolute value of a negative number is its opposite. The absolute value of zero is 0. Use absolute value bars $(|\;|)$ to indicate absolute value.

EXAMPLE

- $|6.32| = 6.32$
- $|-500| = 500$
- $\left|-\frac{8}{9}\right| = \frac{8}{9}$
- $\left|\frac{5}{8} + \frac{3}{8}\right| = |1| = 1$
- $|13| - |-7| = 13 - 7 = 6$
- $-|-100| = -100$
- $|4 + -4| = |0| = 0$
- $|15 - 7| = |8| = 8$

Compute inside the absolute value bars before determining the absolute value.

EXERCISE 4.1

For 1 to 10, evaluate the expression.

1. $|-30|$
2. $|0.5|$
3. $\left|-2\frac{1}{3}\right|$
4. $|-4.8|$
5. $|30|$
6. $|100-40|$
7. $|100|+|-40|$
8. $|0|$
9. $-|-30|$
10. $-|30|$

For 11 to 20, determine whether the statement is true or false.

11. $|-100| < |-1000|$
12. $|50| = |-50|$
13. $|0| > \left|-\frac{1}{2}\right|$
14. $|-8| \leq |8|$
15. $|-3| \leq -3$
16. $|10|+|-4| = |10-4|$
17. $-|20| = -|-20|$
18. $\left|\frac{2}{9}\right| > \left|-\frac{7}{9}\right|$
19. $|6+-6| = |6|+|-6|$
20. $-|9| \leq -|-12|$

Properties of Absolute Value

For any real numbers x and y, the following properties hold:

- $|x| \geq 0$ The absolute value is always nonnegative.
- $|x| = |-x|$ A number and its opposite have the same absolute value.
- $|xy| = |x||y|$ The absolute value of a product equals the product of the absolute values.
- $\left|\frac{x}{y}\right| = \frac{|x|}{|y|}, y \neq 0$ The absolute value of a quotient equals the quotient of the absolute values.
- $|x+y| \leq |x|+|y|$ The absolute value of a sum is less than or equal to the sum of the absolute values.

EXAMPLE

- $|8 \cdot -6| = |8||-6| = 8 \cdot 6 = 48$
- $\left|\frac{-7}{-10}\right| = \frac{|-7|}{|-10|} = \frac{7}{10}$
- $|25 + -25| \leq |25| + |-25|$
- $|-8.35| = 8.35$, which is nonnegative

EXERCISE 4.2

For 1 to 15, determine whether the statement is true or false.

1. $|-30| < 0$

2. $\left|\frac{-10}{2}\right| = \frac{|-10|}{|2|}$

3. $|(-2.5)(-3)| = |-2.5||-3|$

4. $|18 + -2| \leq |18| + |-2|$

5. $|30 \cdot 5| = |30||5|$

6. $|-45| \neq |45|$

7. $-|-100| = -|100|$

8. $\frac{|50|}{|-2|} = \left|\frac{50}{-2}\right|$

9. $|-30||20| = |(-30)(20)|$

10. $|0.39| \geq 0$

11. $|-x| < |x|$

12. $-|-x| < |x|$

13. $|(-a)(-b)| = |-a||-b|$

14. $|-x + y| \leq |-x| + |y|$

15. $\left|\left(-\frac{3}{4}\right)\left(\frac{2}{3}\right)\right| = \left|\left(\frac{3}{4}\right)\left(-\frac{2}{3}\right)\right|$

For 16 to 20, evaluate the expression.

16. $|10| + |-4|$

17. $-|-20|$

18. $\left|\frac{-12}{4}\right|$

19. $|60| + |-60|$

20. $\left|\frac{180}{-3}\right|$

CHAPTER 5

Performing Operations with Real Numbers

(Algebraic) Addition and Subtraction of Real Numbers

Rule 1. To add two nonzero numbers that have the same sign, add their absolute values, and then affix their common sign to the result.

Rule 2. To add two nonzero numbers that have opposite signs, subtract the lesser absolute value from the greater absolute value, and then affix the sign of the number with the greater absolute value to the result; if the two numbers have the same absolute value, their sum is 0.

Rule 3. The sum of 0 and any number is the number.

Rule 4. To subtract two numbers, add the opposite of the second number to the first number using rule 1, 2, or 3.

EXAMPLE

- $-35 + -60 = -95$
- $7.8 + 5.9 = 13.7$
- $-45 + 50 = 5$

- $\frac{2}{9} + -\frac{7}{9} = -\frac{5}{9}$
- $-990.36 + 0 = -990.36$
- $-100 + 100 = 0$
- $-7 - 3 = \underbrace{-7 + -3}_{\text{Do this step mentally.}} = -10$
- $20 - 8 = \underbrace{20 + -8}_{\text{Do this step mentally.}} = 12$
- $-3 + 12 + 5 = 14$
- $-6 + -2 + -10 = -18$
- $15 - (-7) = 15 + 7 = 22$
- $-6 - 2 - 10 = -18$

EXERCISE 5.1

For 1 to 15, complete the statement.

1. $-3 + -6 =$ ________

2. $11 + 23 =$ ________

3. $-18 + 12 =$ ________

4. $-100 + 250 =$ ________

5. $-6 + 0 =$ ________

6. $-78 + 78 =$ ________

7. $-2.5 + 3.25 =$ ________

8. $-\frac{1}{2} + \frac{3}{8} =$ ________

9. $-7 - 9 + 25 - 3 =$ ________

10. $0.08 + 2.12 + 0 - 3.2 =$ ________

11. $25.5 - \frac{3}{4} - 20 =$ ________

12. $|100 - 400| =$ ________

13. $|100| + |-400| =$ ________

14. $5 + 8 + 6 - 2 + 7 - 3 =$ ________

15. $12 + 9 - 6 + 6 + 1 - 7 =$ ________

For 16 to 20, determine the answer by using signed numbers.

16. A certain stock closed on Monday at a selling price of 45.75. Find the closing price of the stock on Friday if it gained 3.87 points on Tuesday, lost 2.5 points on Wednesday, lost 3.62 points on Thursday, and gained 1.45 points on Friday.

17. A *first down* in football is a net gain of 10 or more yards. A football team gained 4 yards on the first play, lost 3 yards on the second play, gained 5 yards on the third play, and gained 6 yards on the fourth play. Did the team make a first down?

18. How far below the surface of the water is the top of a submerged mountain if the ocean floor depth is 12,500 feet and the mountain has a height of 10,190 feet?

19. At 8 a.m., the temperature was $-5°$ F. If the temperature rose 8 degrees in the next hour, what was the temperature reading at 9 a.m.?

20. The *ground speed* of a plane is its horizontal speed relative to the ground. The *air speed* is the speed of the plane in still air. A *headwind* decreases the air speed because it blows in the opposite direction in which a plane is flying. What is the ground speed of a plane meeting a headwind of 35 miles per hour, when the plane's air speed is 415 miles per hour?

(Algebraic) Multiplication and Division Rules for Real (Signed) Numbers

Rule 5. To multiply (or divide) two nonzero numbers that have the same sign, multiply (or divide) their absolute values and keep the product positive.

Rule 6. To multiply (or divide) two nonzero numbers that have opposite signs, multiply (or divide) their absolute values and make the product negative.

Rule 7. Zero divided by any nonzero number is 0.

Rule 8. Division by 0 is undefined.

EXAMPLE

- $(-3)(-40) = 120$
- $19 \cdot 5 = 95$
- $-\frac{3}{4} \cdot \frac{4}{5} = -\frac{3}{\cancel{4}} \cdot \frac{\cancel{4}}{5} = -\frac{3}{5}$
- $\frac{58.2}{-0.3} = -194$
- $\frac{0}{-8} = 0$
- $\frac{70}{0} = \text{not defined}$
- $(-2)(-1)(-3)(-4) = 24$
- $(-2)(-1)(3)(-6) = 36$
- $\frac{-120}{-3} = 40$

In algebra, division is commonly indicated by the fraction bar.

EXERCISE 5.2

For 1 to 15, complete the statement.

1. $(-3)(9) =$ ________
2. $11 \cdot 12 =$ ________
3. $\frac{-18}{-2} =$ ________
4. $-100 \cdot -25 =$ ________
5. $-16 \cdot 0 =$ ________
6. $\frac{78}{13} =$ ________
7. $(0.25)(-400) =$ ________
8. $\left(-\frac{1}{2}\right)\left(\frac{4}{3}\right) =$ ________
9. $(56)\left(-\frac{5}{8}\right) =$ ________
10. $(-6)(-5)(1.5) =$ ________
11. $\frac{3/5}{-6/5} =$ ________
12. $\left|\frac{-50}{2}\right| =$ ________
13. $\left|\frac{-400}{-100}\right| =$ ________
14. $\frac{0}{89} =$ ________
15. $\frac{99}{0} =$ ________

For 16 to 20, determine the answer by using signed numbers.

16. During the past six weeks, Shasta's baby brother gained an average of 6 ounces per week. How many total ounces did Shasta's brother gain?
17. Suppose on a sunny day the temperature decreases 5.4°F for each 1,000-foot rise in elevation. If the temperature at the base of a 3,000-foot mountain is 27°F, what is the temperature at the mountain's summit?
18. A football team gained an average of 3.4 yards in 4 plays. How many total yards did the team make in the 4 plays?
19. The Fahrenheit temperature readings taken at 2-hour intervals on a winter day were $8°, 6°, 2°, 0°, -1°, -4°, -7°$, and $-9°$. What is the average temperature of the 8 readings?
20. The net changes of a certain stock during a 5-day period were -1.63, 2.37, 1.00, -0.12, and 0.87. What was the average net change?

CHAPTER 6

Exponents

An **exponent** is a small raised number written to the upper right of a quantity, called the **base** for the exponent. This representation is an **exponential expression**.

Positive Integer Exponents

If n is a positive integer, then $x^n = \underbrace{x \cdot x \cdot x \cdot \cdots \cdot x}_{n \text{ factors of } x}$. A positive integer exponent tells you how many times to use the base as a factor. For example, $3 \cdot 3 \cdot 3 \cdot 3 \cdot 3$ is written as 3^5. Most commonly, the exponential expression 3^5 is read as "three to the fifth." Other ways you might read 3^5 are "three to the fifth power" or "three raised to the fifth power." For the exponents 2 and 3, the second power of a number is the **square** of the number and the third power of a number is the **cube** of the number.

EXAMPLE

- $8 \cdot 8 = 8^2$ — 8 squared
- $7 \cdot 7 \cdot 7 = 7^3$ — 7 cubed
- $-9 \cdot -9 = (-9)^2$ — -9 squared
- $2 \cdot 2 \cdot 2 \cdot 2 \cdot 2 \cdot 2 \cdot 2 \cdot 2 = 2^8$ — 2 to the eighth power

To evaluate a positive integer exponential expression, perform the indicated multiplication.

To evaluate an expression such as -9^2, square 9 first, and then take the opposite. The exponent applies only to the 9.

To evaluate an expression such as $2 \cdot 5^3$, cube 5 first, and then multiply by 2. The exponent applies only to the 5.

EXAMPLE

- $10^2 = 10 \cdot 10 = 100$
- $4^3 = 4 \cdot 4 \cdot 4 = 64$
- $(-5)^4 = -5 \cdot -5 \cdot -5 \cdot -5 = 625$
- $-9^2 = -(9 \cdot 9) = -81$
- $2 \cdot 5^3 = 2 \cdot 125 = 250$

EXERCISE 6.1

For questions 1 to 5, write the indicated product in exponential form.

1. $3 \cdot 3 \cdot 3 \cdot 3$

2. $\left(-\frac{1}{2}\right)\left(-\frac{1}{2}\right)\left(-\frac{1}{2}\right)\left(-\frac{1}{2}\right)\left(-\frac{1}{2}\right)\left(-\frac{1}{2}\right)\left(-\frac{1}{2}\right)$

3. $(2.5)(2.5)$

4. $(-6)(-6)(-6)$

5. $2 \cdot 2 \cdot 2 \cdot 2 \cdot 2$

For questions 6 to 20, evaluate the expression.

6. $(-4)^2$

7. $\left(\frac{1}{3}\right)^4$

8. $(0.8)^2$

9. $\left(-\frac{4}{5}\right)^3$

10. -3^4

11. $\frac{1}{3}(-12)^2$

12. $(2 + 3)^2$

13. $2^2 + 3^2$

14. $5(-2)^3 + 4(3)^2$

15. $(0.5)^3$

16. $-(-2)^5$

17. $8^2 + 6^2$

18. $(8 + 6)^2$

19. $\left(\frac{10}{13}\right)^2$

20. $\frac{5^4}{100}$

A zero exponent on a nonzero number tells you to put 1 as the answer when you evaluate.

Zero and Negative Integer Exponents

Zero Exponent. $x^0 = 1$ provided $x \neq 0$. (0^0 is undefined; it has no meaning. $0^0 \neq 0$.)

EXAMPLE

- $5^{\circ} = 1$
- $(-300)^0 = 1$
- $\left(\frac{7}{8}\right)^0 = 1$

Negative Integer Exponent. If x is a nonzero real number and n is a positive integer, then $x^{-n} = \frac{1}{x^n}$. A negative integer exponent on a nonzero number tells you to obtain the reciprocal of the exponential expression that has the corresponding positive integer exponent.

It is important that you know the following: $x^{-n} \neq -\frac{1}{x^n}$; $x^{-n} \neq -x^n$.

Rules for Negative Exponents

If x and y are nonzero real numbers and m and n are integers, then the following rules hold:

$$\frac{1}{x^{-n}} = x^n \qquad \left(\frac{x}{y}\right)^{-n} = \left(\frac{y}{x}\right)^n \qquad \frac{x^{-m}}{y^{-n}} = \frac{y^n}{x^m}$$

EXAMPLE

- $10^{-2} = \frac{1}{10^2} = \frac{1}{100}$
- $(4)^{-3} = \frac{1}{4^3} = \frac{1}{64}$
- $(-5)^{-4} = \frac{1}{(-5)^4} = \frac{1}{625}$
- $\left(\frac{1}{0.6}\right)^{-2} = \left(\frac{0.6}{1}\right)^2 = \frac{0.36}{1} = 0.36$
- $\left(\frac{5}{4}\right)^{-3} = \left(\frac{4}{5}\right)^3 = \frac{64}{125}$

EXERCISE 6.2

Evaluate each of the following expressions.

1. $(-1{,}000)^0$

2. $\left(\sqrt{30}\right)^0$

3. $(0.8)^0$

4. $(6^{20} + 10^5)^0$

5. $(3 - 3)^0$

6. $(-4 - 7)^0$

7. $\left(\frac{1}{2{,}000}\right)^0$

8. π^0

9. -5^0

10. $(1 - 85^4)^0$

11. 15^{-2}

12. $\left(\frac{3}{4}\right)^{-2}$

13. $(0.25)^{-1}$

14. $\frac{1}{5^{-2}}$

15. $\left(\frac{5}{3}\right)^{-4}$

16. $\frac{1}{(-2)^{-5}}$

17. $(1 - 0.4)^{-2}$

18. $\left(-\frac{2}{3}\right)^{-3}$

19. $-\left(-\frac{3}{7}\right)^{-1}$

20. $(-100 - 200)^{-2}$

Unit Fraction and Rational Exponents

If n is even, both $x^{1/n}$ and $\sqrt[n]{x}$ are equal to the positive nth root of x.

Unit Fraction Exponent. If x is a real number and n is a positive integer, then $x^{1/n} = \sqrt[n]{x}$, provided that when n is even, $x \geq 0$. (See Lesson 1.2 for a discussion of the meaning of $\sqrt[n]{x}$.)

EXAMPLE

- $25^{1/2} = \sqrt{25}$ (Do this step mentally) $= 5$
- $(-64)^{1/3} = -4$
- $\left(\frac{32}{243}\right)^{1/5} = \frac{2}{3}$
- $(-16)^{1/2}$ is not a real number because it is an even root of a negative number.

Rational Exponent. If x is a real number and m and n are positive integers, then $x^{m/n} = (x^{1/n})^m = (x^m)^{1/n}$, provided that $x^{1/n}$ is a real number. To evaluate $x^{m/n}$, it is usually more practical to find the nth root of x (provided it is a real number), and then raise the result to the mth power.

EXAMPLE

- $1{,}000^{2/3} = \left(1{,}000^{1/3}\right)^2 = 10^2 = 100$
- $(-32)^{3/5} = \left((-32)^{1/5}\right)^3 = (-2)^3 = -8$
- $0.64^{3/2} = \left(0.64^{1/2}\right)^3 = 0.8^3 = 0.512$

EXERCISE 6.3

For 1 to 10, write the radical expression as an exponential expression.

1. $\sqrt{100}$
2. $\sqrt{625}$
3. $\sqrt[3]{-8}$
4. $\left(\sqrt[3]{27}\right)^4$
5. $\left(\sqrt[4]{16}\right)^5$
6. $\left(\sqrt[3]{1{,}000}\right)^2$
7. $\left(\sqrt{9}\right)^4$
8. $\left(\sqrt{5}\right)^2$
9. $\left(\sqrt{\frac{4}{9}}\right)^3$
10. $\left(\sqrt[3]{-0.008}\right)^5$

For 11 to 20, evaluate the expression.

11. $100^{1/2}$
12. $625^{1/2}$
13. $(-8)^{1/3}$
14. $-(27)^{4/3}$
15. $(16)^{5/4}$
16. $(1{,}000)^{2/3}$
17. $-(9)^{4/2}$
18. $(5)^{2/2}$
19. $\left(\frac{4}{9}\right)^{3/2}$
20. $(-0.008)^{5/3}$

Rules for Exponents

For all real numbers x and y and rational numbers m, n, and p, the following rules hold provided that all the roots are real numbers and no denominator is zero.

Product Rule for Same Base Exponential Expressions. $x^m x^n = x^{m+n}$

Quotient Rule for Same Base Exponential Expressions. $\dfrac{x^m}{x^n} = x^{m-n}$

Rule for a Power to a Power. $(x^m)^p = x^{mp}$

Rule for the Power of a Product. $(xy)^p = x^p y^p$

Rule for the Power of a Quotient. $\left(\dfrac{x}{y}\right)^p = \dfrac{x^p}{y^p}$

EXAMPLE

- $x^2x^3 = x^{2+3} = x^5$
- $\dfrac{x^5}{x^3} = x^{5-3} = x^2$
- $\left(\dfrac{x^{-5}}{x^3}\right)^{-2} = (x^{-5-3})^{-2} = (x^{-8})^{-2} = x^{16}$
- $(y^3)^5 = y^{3\cdot 5} = y^{15}$
- $(2y^2z)^4 = (2)^4(y^2)^4(z)^4 = 16y^8z^4$
- $\left(\dfrac{-4x}{5y}\right)^3 = \dfrac{(-4x)^3}{(5y)^3} = \dfrac{(-4)^3(x)^3}{(5)^3(y)^3} = \dfrac{-64x^3}{125y^3} = -\dfrac{64x^3}{125y^3}$

EXERCISE 6.4

Apply the rules for exponents. Write the answer so that all exponents are positive. Assume all variables are positive real numbers.

1. z^7z^3
2. $\dfrac{y^5}{y^2}$
3. $(x^6)^2$
4. $\left(\dfrac{a}{b}\right)^5$
5. $(xyz)^8$
6. $(x^5y^3z)(x^2yz^2)$
7. $(x^{-2}y^6z^{-1})(x^2y^{-4}z^{-2})$
8. $(xz^7)(xyz^{-3})$

9. $\dfrac{y^2}{y^{-3}}$

10. $(9x^6)^2$

11. $\left(\dfrac{2a}{b}\right)^5$

12. $(2xyz)^5$

13. $\dfrac{x^5y^{-3}}{x^3y}$

14. $\left(x^{1/2}y^{1/4}\right)^4$

15. $\left(\dfrac{x^{-5}}{3y^{-1}}\right)^{-2}$

16. $\left(\dfrac{2}{5}\right)^4$

17. $\left(\dfrac{9}{25}\right)^{-1/2}$

18. $(a^{-2}b^4c^{-1})(a^{-5}b^3c^4)$

19. $(16x^{16}y^8)^{1/4}$

20. $(0.027c^{12}d^{15})^{1/3}$

CHAPTER 7

Radicals

Expressing Rational Exponents as Radicals

If x is a real number and m and n are positive integers for which $\sqrt[n]{x}$ is a real number, then $x^{m/n}$ can be expressed as $\left(\sqrt[n]{x}\right)^m$ or $\sqrt[n]{x^m}$, whichever one is more suitable for the situation.

EXAMPLE

- $8^{2/3} = \left(\sqrt[3]{8}\right)^2 = (2)^2 = 4$ or $8^{2/3} = \sqrt[3]{8^2} = \sqrt[3]{64} = 4$
- $(5x)^{3/4} = \left(\sqrt[4]{5x}\right)^3$ or $(5x)^{3/4} = \sqrt[4]{(5x)^3} = \sqrt[4]{125x^3}, x \geq 0$
- $4^{2/3} = \left(\sqrt[3]{4}\right)^2$ or $4^{2/3} = \sqrt[3]{4^2} = \sqrt[3]{16}$

EXERCISE 7.1

Use $x^{m/n} = \left(\sqrt[n]{x}\right)^m$ to write the expression in radical form and simplify, if possible. Assume all variables are positive real numbers.

Commonly, *simplify* means to transform an expression so that it is less complex and easier to use.

1. $8^{5/3}$

2. $16^{3/4}$

3. $(8x^6y^3)^{5/3}$

4. $\left(\dfrac{16z^4}{49x^8y^2}\right)^{1/2}$

5. $(125r^9s^{21})^{1/3}$

6. $(81x^4y^{12}c^{16})^{3/4}$

7. $(a^{12})^{-5/3}$

8. $(-8a^3b^6c^9)^{1/3}$

9. $-(100m^4n^{12})^{1/2}$

10. $-(x^9y^{15})^{4/3}$

11. $(-8x^9y^{15})^{4/3}$

12. $-(625r^4s^8)^{1/4}$

13. $(289x^6y^{14})^{1/2}$

14. $(169x^2y^2)^{1/2}$

15. $(256a^{12}b^8)^{3/4}$

16. $(1{,}000x^3y^6)^{2/3}$

17. $(-27x^3y^6z^3)^{2/3}$

18. $(0.008r^3s^{12})^{1/3}$

19. $(1.44a^2b^{10}c^4)^{1/2}$

20. $(0.04x^{40})^{5/2}$

Product and Quotient Rules for Radicals

If n is a positive integer and x and y are real numbers, the following rules hold provided that all the roots are real numbers and no denominator is zero.

Product Rule for Radicals. $\sqrt[n]{xy} = \sqrt[n]{x} \cdot \sqrt[n]{y}$

Quotient Rule for Radicals. $\sqrt[n]{\dfrac{x}{y}} = \dfrac{\sqrt[n]{x}}{\sqrt[n]{y}}$

EXAMPLE

- $\sqrt{150} = \sqrt{25}\sqrt{6} = 5\sqrt{6}$
- $2\sqrt{150} = 2\sqrt{25}\sqrt{6} = 2 \cdot 5\sqrt{6} = 10\sqrt{6}$
- $\sqrt[3]{250x^7} = \sqrt[3]{125x^6} \cdot \sqrt[3]{2x} = 5x^2\sqrt[3]{2x}$
- $\sqrt[3]{64x^6} = \sqrt[3]{64} \cdot \sqrt[3]{x^6} = 4x^2$
- $\sqrt{\dfrac{7}{25}} = \dfrac{\sqrt{7}}{\sqrt{25}} = \dfrac{\sqrt{7}}{5}$
- $\sqrt[5]{-32y^{15}} = \sqrt[5]{-32} \cdot \sqrt[5]{y^{15}} = -2y^3$

EXERCISE 7.2

Transform the radical expression into a simpler form. Assume all variables are positive real numbers.

1. $\sqrt{32}$
2. $\sqrt{75}$
3. $\sqrt[3]{128}$
4. $\sqrt{200}$
5. $\sqrt{176}$
6. $3\sqrt{150}$
7. $-\sqrt{48}$
8. $\frac{2}{3}\sqrt{108}$
9. $-\frac{2}{9}\sqrt{162}$
10. $\frac{1}{2}\sqrt{40}$
11. $\sqrt{49x^3y^6}$
12. $\sqrt{128a^9b^8c^{10}}$
13. $\frac{2}{5x}\sqrt{75x^4}$
14. $3m\sqrt[3]{8m^7}$
15. $\frac{1}{8x}\sqrt[4]{256x^5y}$
16. $2r\sqrt{80r^7s^2}$
17. $-7xy\sqrt{45x^3y^5}$
18. $\frac{5a}{9b}\sqrt{80a^3b}$
19. $\frac{3x}{2y}\sqrt[5]{64x^{18}y^{12}}$
20. $-5xy\sqrt[3]{-8x^5y^3}$

Transforming Radicals into Simplified Form

A radical of index n is in **simplified form** if it has (1) no fractions in the radicand, (2) no radicals in a denominator, and (3) no factor in the radicand that is a perfect nth power.

EXAMPLE

$$\sqrt[3]{16} = \sqrt[3]{8} \cdot \sqrt[3]{2} = 2\sqrt[3]{2}$$

$$\sqrt{20x^9y^8} = \sqrt{4x^8y^8} \cdot \sqrt{5x} = 2x^4y^4\sqrt{5x},\ x, y \geq 0$$

$$\sqrt{\frac{5}{3}} = \frac{\sqrt{5}}{\sqrt{3}} = \frac{\sqrt{5} \cdot \sqrt{3}}{\sqrt{3} \cdot \sqrt{3}} = \frac{\sqrt{15}}{3}$$

$$\frac{1}{\sqrt{2}} = \frac{1}{\sqrt{2}} \cdot \frac{\sqrt{2}}{\sqrt{2}} = \frac{\sqrt{2}}{\left(\sqrt{2}\right)^2} = \frac{\sqrt{2}}{2}$$

This process of removing radicals from the denominator is called **rationalizing the denominator.**

Multiplying by $\frac{\sqrt{2}}{\sqrt{2}}$ is the same as multiplying by 1, so the value of the expression is not changed.

EXERCISE 7.3

Express the radical expression in simplified form. Assume all variables are positive real numbers.

1. $\sqrt{\frac{1}{2}}$
2. $\sqrt{\frac{1}{3}}$
3. $\sqrt{\frac{1}{4}}$
4. $\sqrt{\frac{1}{5}}$
5. $\sqrt{\frac{5}{6}}$
6. $\sqrt{\frac{3}{7}}$
7. $\sqrt{\frac{7}{9}}$
8. $10\sqrt{\frac{3}{10}}$
9. $\sqrt[3]{\frac{3}{4}}$
10. $\sqrt{\frac{x}{y}}$
11. $\sqrt{\frac{2a}{3b}}$
12. $\sqrt{\frac{3x^7}{8x^3y^5}}$
13. $-\sqrt{\frac{y}{3}}$
14. $\frac{5xy}{z^2}\sqrt{\frac{3z^4}{5xy^2}}$
15. $\sqrt[3]{\frac{27a^4}{2b^2c}}$
16. $\frac{3x}{y}\sqrt[3]{\frac{4y^4}{x}}$
17. $\frac{5m}{4n}\sqrt[3]{-\frac{64n^3}{25m^2}}$
18. $-4x\sqrt[3]{-\frac{7y^2}{16x}}$
19. $\frac{7}{16}\sqrt[3]{\frac{8}{49}}$
20. $\sqrt[3]{-\frac{9x^3}{16y^2}}$

CHAPTER 8

Order of Operations

Grouping Symbols

Grouping symbols such as parentheses (), brackets [], and braces {} are used to keep things together that belong together. Fraction bars, absolute value bars | |, and square root symbols $\sqrt{\ }$ are also grouping symbols. As a general rule, when evaluating expressions, do operations in grouping symbols first—especially if you have addition or subtraction inside the grouping symbol.

Do keep in mind that parentheses are also used to indicate multiplication as in (−5)(−8) or for clarity as in −(−35).

EXAMPLE

- $(-8+4)-(5-2)=-4-(3)=-4-3=-7$
- $\dfrac{4-10}{-3-(-6)}=\dfrac{4-10}{-3+6}=\dfrac{-6}{3}=-2$
- $\sqrt{36+64}=\sqrt{100}=10$
- $3(10+5)=3\cdot 15=45$
- $(-8\cdot 4)-(5\cdot -2)=-32-(-10)=-32+10=-22$
- $13-\left|5-(17-8)\right|=13-\left|5-(9)\right|=13-\left|-4\right|=13-\left|-4\right|=13-4=9$

Omit the grouping symbol when it is no longer needed.

Evaluate within the innermost grouping symbols first.

EXERCISE 8.1

Evaluate the numerical expression.

1. $(-5 \cdot -4) - (4 \cdot -2)$
2. $\left(\frac{1}{2} \cdot 4\right) + (8 \cdot 3) - (9 \cdot 5)$
3. $\frac{5 - 16}{-4 - 7}$
4. $\frac{8 + 2}{-14 + 19} + \frac{24 - 36}{-4}$
5. $15 - |10 - 24|$
6. $|20 - 30| + 10$
7. $\frac{1}{2}(-3 - 5)$
8. $\sqrt{16 + 9}$
9. $7(8 - 10)$
10. $(-2 + 5)(8 - 7)$
11. $6 - |5 - 9|$
12. $3 - (4 \cdot 8)$
13. $5 - 2[6 - (5 \cdot 2)]$
14. $2\sqrt{100 - 36}$
15. $4 - 5\sqrt{60 + 4}$
16. $\frac{4 + 8}{2 - 5} + \left(\frac{12}{25}\right)\left(\frac{75}{6}\right)$
17. $(5 + 7)\frac{1}{3}$
18. $2 - (4 \cdot 5) - 6(3 - 4)$
19. $1 - [(5 \cdot 3) - 3 + 2]$
20. $10 - 2[7 - (2 \cdot 3) - (8 - 3)]$

Order of Operations

A **numerical expression** is a meaningful combination of numbers using the ordinary operations of arithmetic. Its **value** is the real number that results from performing the indicated computations.

Follow the order of operations to evaluate numerical expressions. Use the mnemonic "Please Excuse My Dear Aunt Sally"—abbreviated as PE(MD)(AS)—to help you remember the following order:

1. Do computations inside **P**arentheses (or other grouping symbols).
2. Evaluate **E**xponential expressions (also, evaluate absolute value, square root, and other root expressions).
3. Perform **M**ultiplication and **D**ivision, in the order in which these operations occur from left to right.
4. Perform **A**ddition and **S**ubtraction, in the order in which these operations occur from left to right.

Multiplication does not always have to be done before division, or addition before subtraction. You multiply and divide in the order they occur in the problem. Similarly, you add and subtract in the order they occur in the problem.

EXAMPLE

$6 - 5(9 - 7)^3 = 6 - 5(2)^3 = 6 - 5(8) = 6 - 40 = -36$

$10 - 4(3 - 4 \cdot 2)^2 = 10 - 4(3 - 8)^2 = 10 - 4(-5)^2 = 10 - 4(25)$
$= 10 - 100 = -90$

$\dfrac{60}{(8+4)} - 3 \cdot 2^2 + (3+1)^3 = \dfrac{60}{12} - 3 \cdot 2^2 + (4)^3 = \dfrac{60}{12} - 3 \cdot 4 + 64$
$= 5 - 12 + 64 = 57$

Follow the order of operations within grouping symbols.

EXERCISE 8.2

Evaluate the numerical expression.

1. $9 - 4(20 - 17)^2$

2. $\dfrac{1}{2} \cdot 4 + 8 \cdot 3 - 9 \cdot 5$

3. $100 + 8 \cdot 3^2 - 63 + 2(1 + 5)$

4. $\dfrac{-7 + 25}{-3} + |8 - 15| - (5 - 3)^3$

5. $15 - \left|\sqrt{100} - 24\right|$

6. $|200 - 300| + 10^2$

7. $\dfrac{1}{2}(-3 - 5)^2$

8. $-2\sqrt{16 + 9}$

9. $\dfrac{7(8 - 10)}{14}$

10. $(-2 + 5)^2 + (8 - 7)^3$

11. $\sqrt{36} - |5^2 - 9|$

12. $3 \cdot 5 - 4 \cdot 8$

13. $5 - 2\left[6 - (5 \cdot 2^3 + 4)\right]$

14. $2(100 - 36)^{1/2}$

15. $2\left(100^{1/2} - 36^{1/2}\right)$

16. $2\left(\dfrac{5 + 9}{-2 - 5}\right) - \left(\dfrac{1}{3}\right)(14 - 2)$

17. $\left(25^{1/2} + 49^{1/2}\right)\dfrac{1}{3}$

18. $2 - 4 \cdot 5 - 6(5 - 4)$

19. $1 - (5 \cdot 3 - 3 + 2)$

20. $10 - 2\left[7 - 2 \cdot 3 - (8 - 3)\right]$

CHAPTER 9

Algebraic Expressions and Formulas

Algebraic Expressions

An **algebraic expression** (or, simply, **expression**) is a meaningful combination of constants and one or more variables using the ordinary operations of arithmetic. It is a symbolic representation of a number. In algebraic expressions, writing numbers and variables or two or more variables (with or without numbers) side by side with no multiplication symbol in between indicates multiplication. Thus, $2xyz$ means 2 times x times y times z.

If you are given numerical values for the variables, you can evaluate an algebraic expression by substituting the given numerical value for each variable, and then performing the indicated operations, being sure to *follow the order of operations* as you proceed.

A **constant** is a fixed number such as 10.

Recall that a **variable** is a letter used as a placeholder for a number—or numbers, depending on the situation.

EXAMPLE

- Evaluate $2xyz$ when $x = 3$, $y = -5$, and $z = -2$.
- $2xyz = 2(3)(-5)(-2) = 60$

Watch your signs! It's not uncommon to make careless errors when evaluating negative numbers raised to powers.

EXAMPLE

Evaluate $-2x^5 + 5x^4 - 3x^3 - 7x^2 + x + 4$ when $x = -1$.

$$\begin{aligned} -2x^5 + 5x^4 - 3x^3 - 7x^2 + x + 4 &= -2(-1)^5 + 5(-1)^4 - 3(-1)^3 - 7(-1)^2 \\ &\quad + (-1) + 4 \\ &= -2(-1) + 5(1) - 3(-1) - 7(1) \\ &\quad + (-1) + 4 \\ &= 2 + 5 + 3 - 7 - 1 + 4 \\ &= 6 \end{aligned}$$

EXAMPLE

Evaluate $\sqrt{b^2 - 4ac}$ when $a = 2$, $b = -5$, and $c = -12$

$$\sqrt{b^2 - 4ac} = \sqrt{(-5)^2 - 4(2)(-12)} = \sqrt{25 + 96} = \sqrt{121} = 11$$

EXERCISE 9.1

Find the value of the expression if $x = 5$, $y = 4$, and $z = -2$.

1. $3(x + y)$
2. $x + y(y - z)$
3. $(x + y)(y - z)$
4. $\dfrac{5(3z - y)}{z(5x - 2z + 6)}$
5. $xy(6z - 7)$
6. $xyz(x + y + z)$
7. $2x + \left(y + z^2\right)$
8. $4(2x + y) - x(z - 5y)$
9. $(xy)^2$
10. $2x^2z - 3y^2$
11. $x^3 - z^3$
12. $4z^2 - 10$
13. $x^2 + 6x + 9$
14. $(x - y)(x + y)$
15. $\dfrac{x^3}{z^4}$
16. $\dfrac{9x^2}{75} - \dfrac{4y^2}{64}$
17. $\dfrac{y^{1/2}}{z}$
18. $\dfrac{2(y + z)^2}{5}$
19. $\dfrac{\sqrt{16x - 15y}}{\sqrt{x}}$
20. $\dfrac{x^2 - 4x + 4}{x^2 - 4}$

Formulas

Formulas are rules that model relationships in real-life situations. For example, the formula $C = \frac{5}{9}(F - 32)$ models the relationship between Celsius temperature and Fahrenheit temperature.

Use your skills in evaluating algebraic expressions to evaluate formulas for given numerical values.

EXAMPLE

Given the formula $C = \frac{5}{9}(F - 32)$, find the value of C when $F = 68$.

$C = \frac{5}{9}(F - 32) = \frac{5}{9}(68 - 32) = \frac{5}{9}(36) = 20$

EXAMPLE

Given the formula $A = \frac{1}{2}(b_1 + b_2)h$, find the value of A when $b_1 = 12$, $b_2 = 15$, and $h = 10$.

$A = \frac{1}{2}(b_1 + b_2)h = \frac{1}{2}(12 + 15)(10) = \frac{1}{2}(27)(10) = 135$

EXERCISE 9.2

Answer the following questions.

1. Given the formula $C = \frac{5}{9}(F - 32)$, find the value of C when $F = 212$.
2. Given the formula $A = \frac{1}{2}(b_1 + b_2)h$, find the value of A when $b_1 = 10$, $b_2 = 6$, and $h = 5$.
3. Given the formula $d = rt$, find the value of d when $r = 65$ and $t = 3$.
4. Given the formula $I = \frac{E}{R}$, find the value of I when $E = 220$ and $R = 20$.
5. Given the formula $A = \frac{1}{2}bh$, find the value of A when $b = 14$ and $h = 12$.

CHAPTER 10

Polynomial Terminology

Monomials

A **monomial** is a single **term** that is the product of a constant and one or more variables with nonnegative integer exponents.

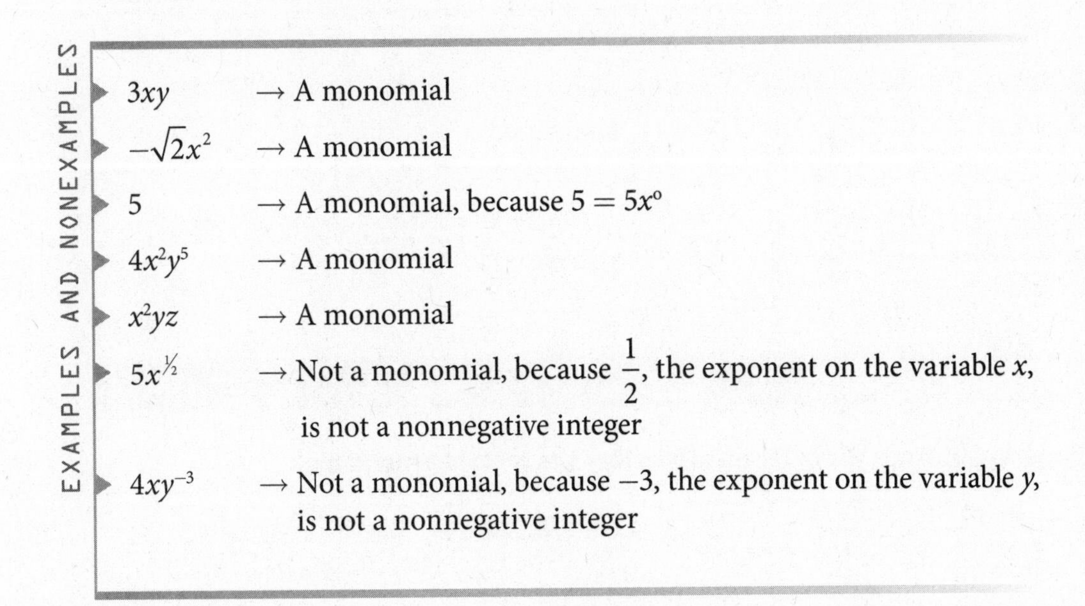

EXAMPLES AND NONEXAMPLES

- $3xy$ $\rightarrow$ A monomial
- $-\sqrt{2}x^2$ $\rightarrow$ A monomial
- 5 $\rightarrow$ A monomial, because $5 = 5x^0$
- $4x^2y^5$ $\rightarrow$ A monomial
- x^2yz $\rightarrow$ A monomial
- $5x^{1/2}$ $\rightarrow$ Not a monomial, because $\frac{1}{2}$, the exponent on the variable x, is not a nonnegative integer
- $4xy^{-3}$ $\rightarrow$ Not a monomial, because -3, the exponent on the variable y, is not a nonnegative integer

The numerical factor of a monomial is its **numerical coefficient**:

EXAMPLE

Monomial	Numerical coefficient
$3xy$	3
$-\sqrt{2}x^2$	$-\sqrt{2}$
$5x$	5
$4x^2y^5$	4
x^2yz	1

If no coefficient is explicitly written, it is understood to be 1.

The sum of the exponents of the variable factors of a monomial is its **degree.** The degree of a nonzero constant monomial is 0. The degree of the zero monomial is undefined:

EXAMPLE

Monomial	Degree
$3xy$	2
$-\sqrt{2}x^2$	2
5	0
$4x^2y^5$	7
0	Undefined
x^2yz	4

EXERCISE 10.1

For 1 to 10, state Yes or No as to whether the term is a monomial.

1. $-8xy^3$

2. $2.5x^2$

3. 0

4. $3a^5b$

5. 27

6. $4x^{-3}y^2$

7. $\sqrt{5}x^3y^2$

8. $-\frac{3}{4}x^5$

9. $50x^{1/3}y^3$

10. $\frac{9x}{y^2}$

For 11 to 20, state (a) the coefficient and (b) the degree of the monomial.

11. $-5.25xy^3$

12. $20x^2$

13. $5x$

14. $13a^5b$

15. $\sqrt{8}y^2x^3z$

16. xy^2z

17. $-0.5x^3y^2$

18. $-\frac{3}{4}a^5b^2c^2$

19. $50x^2y^3z$

20. $\frac{2}{3}mn$

One-Variable Polynomials

A **polynomial** is a single monomial or the (algebraic) sum of two or more monomials. Polynomials with exactly one term, two terms, or three terms are **monomials, binomials,** or **trinomials**, respectively.

A one-variable polynomial in x is defined as follows:

$a_nx^n + a_{n-1}x^{n-1} + a_{n-2}x^{n-2} + \ldots + a_1x^1 + a_0$*, where $a_n, a_{n-1}, \ldots, a_1$ and a_0** are constant coefficients, n is a nonnegative integer, and x is a variable. If $a_n \neq 0$, the **leading term** is a_nx^n, the term that contains the highest power of the variable, and the **leading coefficient** is a_n. The last term, a_0, is the **constant term**. The **degree of the polynomial** is n, the highest power of the variable in the polynomial. The degree of a monomial consisting of a single nonzero constant is zero. The degree of the zero polynomial is undefined.

Normally, polynomials of four or more terms are not given special names based on the number of terms.

* It is customary to write polynomials in descending powers of the variable.

** The notation a_n is read "a sub n."

EXAMPLE

- The polynomial $\frac{1}{2}x^4 + x^2 - \frac{2}{5}$ is a trinomial with leading coefficient $\frac{1}{2}$ and degree 4.
- The polynomial $-10y$ is a monomial with leading coefficient -10 and degree 1.
- The polynomial $-16t^2 + 50$ is a binomial with leading coefficient -16 and degree 2.

EXERCISE 10.2

For each polynomial (a) state the number of terms, (b) give the leading coefficient, and (c) state the degree.

1. $-7x^3 + 5x^2 - 4x + 1$

2. $4x^2 - 1$

3. $18 - 3x + 5x^4$

4. $-5x^5 + 3x^4 + 2x^2 - 4x + 1$

5. $4x^2 - 4x + 1$

6. $7x^5 + 3x^4 - 3x^3 - x^2 + 7x - 5$

7. $x^3 - 3x^2 + 3x - 1$

8. $x^2 - 6x + 9$

9. $2x^2 - 5x - 3$

10. $20x$

CHAPTER 11

Adding and Subtracting Polynomials

Adding and Subtracting Monomials

Monomials that are constants or that have exactly the same variable factors (that is, same letters with the same corresponding exponents) are **like terms**. Like terms are the same except, perhaps, for their coefficients. Monomials that are not like terms are **unlike terms**.

Use the distributive property to add (or subtract) like terms. The sum (or difference) of unlike terms can only be indicated.

EXAMPLE

- $-10x + 25x = \underbrace{(-10 + 25)x}_{\text{Do this step mentally.}} = 15x$
- $4x^2 - 7x^2 = -3x^2$
- $-5xy - 2xy = -7xy$
- $25 + 25x = 25 + 25x$ These are not like terms, so their sum is only indicated.
- $x + 2x = 3x$
- $3(x + 1) + 2(x + 1) = 5(x + 1)$
- $-2xy^2 + 5xy^2 - xy^2 = 2xy^2$

EXERCISE 11.1

Add or subtract as indicated.

1. $6x + 2x$

2. $-4x^3 - 3x^3$

3. $3x - 3x$

4. $-5x^4 + 2x^3$

5. $-5z^2 + 4z^2$

6. $x - x$

7. $-9x^3y^2 - x^3y^2$

8. $x^4 - 3x^4$

9. $18 - 20$

10. $2.5x + 3.5x$

11. $5x - 9x + 2x$

12. $-10x^3 - 3x^3 + 14x^3$

13. $\frac{1}{2}x - \frac{3}{4}x$

14. $6(x + 5) - 2(x + 5) + 4(x + 5)$

15. $-1.5z^2 + 4.3z^2 + 2.1z^2$

For 16 to 20, determine the answer in terms of the given variable or variables.

16. Find the sum of x^2, $-3x^2$, and $9x^2$.

17. Combine $-9y$, $-y$, $-5y$, and $2y$.

18. Find the sum of $8x$, $-2y$, $-5y$, and $6x$.

19. Find the perimeter of a triangle whose sides measure $2.1x$, $4.0x$, and $3.5x$.

20. Find the perimeter of a quadrilateral whose sides measure $\frac{1}{2}x, \frac{3}{8}y, \frac{1}{4}x$, and $\frac{5}{8}y$.

Adding and Subtracting Polynomials

To add or subtract polynomials, combine like terms and indicate addition or subtraction of unlike terms.

Rearranging so that like terms are together can be done mentally. However, writing out this step helps you avoid careless errors.

Think of the minus sign between the two polynomials as "+− 1", so multiply each term inside the second parentheses by − 1.

EXAMPLE

$$\begin{aligned}(9x^2 - 6x + 2) + (-7x^2 - 5x + 3) &= 9x^2 - 6x + 2 - 7x^2 - 5x + 3\\ &= 9x^2 - 7x^2 - 6x - 5x + 2 + 3\\ &= 2x^2 - 11x + 5\end{aligned}$$

$$\begin{aligned}(4x^3 + 3x^2 - x + 8) - (8x^3 + 2x - 10) &= 4x^3 + 3x^2 - x + 8 - 8x^3 - 2x + 10\\ &= 4x^3 - 8x^3 + 3x^2 - x - 2x + 8 + 10\\ &= -4x^3 + 3x^2 - 3x + 18\end{aligned}$$

EXERCISE 11.2

Add or subtract as indicated.

1. $(12x^3 - 5x^2 + 10x - 60) + (3x^3 - 7x^2 - 1)$
2. $(10x^2 - 5x + 3) + (6x^2 + 5x - 13)$
3. $(20x^3 - 3x^2 - 2x + 5) + (9x^3 + x^2 + 2x - 15)$
4. $(10x^2 - 5x + 3) - (6x^2 + 5x - 13)$
5. $(20x^3 - 3x^2 - 2x + 5) - (9x^3 + x^2 + 2x - 15)$
6. $(8x^3 - 3x^2 + 6x - 2) + (3x^4 + 2x^3 + x^2 - x)$
7. $(10y^2 - 15y - 3) + (4y^2 + 5y - 13)$
8. $(5x^3 - 4x^2 - 3x + 5) + (6x^3 + 2x^2 - 2x - 15) - (11x^3 - 5x^2 + 2x - 5)$
9. $(5x^2 - 10x - 3) - (x^2 - 5x + 10) + (x^3 - 3x^2 - 2x + 1)$
10. $(-2x^4 + 3x^3 - 2x + 5) - (7x^3 + 2x - 15) - (x^4 - 4)$

For 11 to 20, determine the answer in terms of the given variable or variables.

11. Find the sum of $2x^2 - 3x + 5$ and $4x^2 + 6x - 3$.
12. Subtract $3a^3 - 4a + 7$ from zero.
13. Subtract $2z^2 + z - 1$ from $z^3 + 2z^2 - z$.
14. Subtract $a^2 - 4a + 6$ from a^2.
15. Subtract $2x^2$ from $x^3 + 2x^2$.
16. Subtract $5x^3 - 4x + 8$ from $x^3 - 3x^2$.
17. Subtract $3x^2 - 3x + 1$ from $5x^4 - 2x^3$.
18. Find the sum of $2x^2 + 4xy + y^2$, $x^2 - y^2$, and $2y^2 - 4xy - x^2$.
19. Subtract $x^2 - 4xy + 4y^2$ from the sum of $x^2 - 2xy + y^2$ and $x^2 + 2xy + y^2$.
20. The sum of the measures of the angles of a triangle is 180 degrees. Find the measure, in degrees, of the third angle of a triangle in which the measure of one angle, in degrees, is $5x + 15$ and the measure of a second angle, in degrees, is $4x - 10$.

CHAPTER 12

Multiplying Polynomials

Multiplying Monomials

Multiply monomials by using the rules for exponents and the commutative and associative properties for real numbers.

EXAMPLE

- $(5x^5)(3x^2) = \underbrace{(5 \cdot 3)(x^5 \cdot x^2)}_{\text{Do this step mentally.}} = 15x^7$
- $(-2a^3b^4)(8ab^2) = -16a^4b^6$
- $(6x)(-2x) = -12x^2$
- $(5)(-4x) = -20x$
- $(4x^2y^5)(-2xy^3)(-3xy) = 24x^4y^9$
- $(10)(x) = 10x$

EXERCISE 12.1

For 1 to 10, multiply as indicated.

1. $(-4x^3)(3x^2)$
2. $(2x^3y^5)(6xy^2)$
3. $(-5x)(-2x)$
4. $(5x)(-4)$
5. $(-2x^2y^3)(-5xy^3)(-xy)$
6. $(-4y^3)(-z)$
7. $\left(\frac{1}{2}m^4n^2\right)(8m^2n)$
8. $(-5xyz)(2xy^2)$
9. $(xy^2)(-2x)(-4x^2y)(3xy)$
10. $(-2a)(-2a)(-2a)(-2a)(-2a)$

For 11 to 15, determine the answer in terms of the given variable or variables.

11. Find the product of $5xy$ and 3.
12. Multiply $9ab^4$ by -1.
13. Multiply $-6x^5y^4z$ by $2x^3z^2$.
14. The length, in feet, of a rectangle is $10x$ and its width, in feet, is $6x$. Find its area. Hint: $A = lw$.
15. How far can a vehicle travel in 3 hours at an average speed, in miles per hour, of $7x$.

Multiplying a Polynomial by a Monomial

To multiply a polynomial by a monomial, multiply each term of the polynomial by the monomial.

EXAMPLE

$$2(x+5) = 2 \cdot x + 2 \cdot 5 = 2x + 10$$

$$x(3x-2) = x \cdot 3x - x \cdot 2 = 3x^2 - 2x$$

$$\begin{aligned} -8a^3b^5(2a^2 - 7ab^2 - 3) &= (-8a^3b^5)(2a^2) - (-8a^3b^5)(7ab^2) - (-8a^3b^5)(3) \\ &= -16a^5b^5 + 56a^4b^7 + 24a^3b^5 \end{aligned}$$

$$\begin{aligned} x^2(2x^4 + 4x^3 - 3x + 6) &= x^2 \cdot 2x^4 + x^2 \cdot 4x^3 - x^2 \cdot 3x + x^2 \cdot 6 \\ &= 2x^6 + 4x^5 - 3x^3 + 6x^2 \end{aligned}$$

EXERCISE 12.2

For 1 to 10, multiply as indicated.

1. $5(x + 3)$

2. $x(4x - 5)$

3. $-2(2x + 3)$

4. $3x(2x - 1)$

5. $a(c + d)$

6. $2x(x - 5)$

7. $5x(2x - 3)$

8. $4(2x - 3)$

9. $-5x^3y^2(2x^2 - 6xy^2 + 3)$

10. $z^3(3x^4z + 4x^3z^2 - 3xz + 5)$

For 11 to 15, determine the answer in terms of the given variable or variables.

11. Find the product of $x + 3$ and x.

12. Multiply $9a + 5b$ by -1.

13. Multiply $-5z^2 + 3z - 2$ by $2z$.

14. The length, in meters, of a rectangle is $2x - 3$ and its width, in meters, is $5x$. Find its area. Hint: $A = lw$.

15. Find the cost of $2x + 7$ dozen if the cost per dozen is x dollars.

Multiplying Binomials

Method 1 To multiply two binomials, multiply all the terms of the second binomial by each term of the first binomial, and then combine like terms.

EXAMPLE

$$(a + b)(c + d) = a \cdot c + a \cdot d + b \cdot c + b \cdot d$$
$$= ac + ad + bc + bd$$

$$(2x + 1)(x + 5) = 2x \cdot x + 2x \cdot 5 + 1 \cdot x + 1 \cdot 5$$
$$= 2x^2 + 10x + x + 5$$
$$= 2x^2 + 11x + 5$$

$$(2x + 3)(x - 5) = 2x \cdot x - 2x \cdot 5 + 3 \cdot x - 3 \cdot 5$$
$$= 2x^2 - 7x - 15$$

$(2x + 1)(x + 5) \neq 2x^2 + 5$. Don't forget about $2x \cdot 5 + 1 \cdot x = 11x$!

Method 2 To multiply two binomials, use FOIL.

From Method 1, you can see that to find the product of two binomials, you compute four products, called **partial products**, using the terms of the two binomials. The FOIL method is a quick way to get those four partial products. FOIL is an acronym for first, outer, inner, and last. Here is how FOIL works for finding the four partial products for $(a + b)(c + d)$.

Be aware that the FOIL method works only for the product of two binomials.

1. Multiply the two **F**irst terms. $a \cdot c$
2. Multiply the two **O**uter terms. $a \cdot d$
3. Multiply the two **I**nner terms. $b \cdot c$
4. Multiply the two **L**ast terms. $b \cdot d$

Forgetting to compute the middle terms is the most common error when one is finding the product of two binomials.

The inner and outer partial products are the **middle terms**.

EXAMPLE

$$(2x+1)(x+5) = \underbrace{2x \cdot x}_{\text{First}} + \underbrace{2x \cdot 5}_{\text{Outer}} + \underbrace{1 \cdot x}_{\text{Inner}} + \underbrace{1 \cdot 5}_{\text{Last}}$$
$$= 2x^2 \underbrace{+10x + 1x}_{\text{middle terms}} + 5$$
$$= 2x^2 + 11x + 5$$

$$(5x+4)(2x-3) = 5x \cdot 2x - 5x \cdot 3 + 4 \cdot 2x - 4 \cdot 3$$
$$= 10x^2 - 15x + 8x - 12$$
$$= 10x^2 - 7x - 12$$

EXERCISE 12.3

For 1 to 10, multiply as indicated.

1. $(x + 3)(x - 2)$
2. $(4x - 3)(4x + 3)$
3. $(2x - y)(x + 2y)$
4. $(5x + 4)(2x - 3)$
5. $(z - 2)(z + 5)$
6. $(x + y)^2$
7. $(a - b)^2$
8. $(x - \sqrt{2})(x + \sqrt{2})$
9. $(\sqrt{7} - \sqrt{3})(\sqrt{7} + \sqrt{3})$
10. $(x^2 - 5)(x^2 + 5)$

For 11 to 15, determine the answer in terms of the given variable or variables.

11. Find the product of $2x + 3$ and $x - 1$.

12. Multiply $6 + y$ by $5 - 2y$.

13. Multiply $(x + a)$ by $(x - a)$.

14. The length, in feet, of a rectangle is $x + 5$ and its width, in feet, is $3x - 1$. Find its area. Hint: $A = lw$.

15. The length, in meters, of the base of a triangular sign is $3x + 4$ with a height, in meters, of $2x + 5$. Find its area. Hint: $A = \frac{1}{2}bh$.

Multiplying Two Polynomials

To multiply two polynomials, multiply all the terms of the second polynomial by each term of the first polynomial, and then combine like terms.

EXAMPLE

$$\begin{aligned}(2x - 1)(3x^2 - 5x + 4) &= 2x \cdot 3x^2 - 2x \cdot 5x + 2x \cdot 4 - 1 \cdot 3x^2 + 1 \cdot 5x - 1 \cdot 4 \\ &= 6x^3 - 10x^2 + 8x - 3x^2 + 5x - 4 \\ &= 6x^3 - 13x^2 + 13x - 4\end{aligned}$$

$$\begin{aligned}(4x^2 + 2x - 5)(2x^2 - x - 3) &= 4x^2 \cdot 2x^2 - 4x^2 \cdot x - 4x^2 \cdot 3 + 2x \cdot 2x^2 \\ &\quad - 2x \cdot x - 2x \cdot 3 - 5 \cdot 2x^2 + 5 \cdot x + 5 \cdot 3 \\ &= 8x^4 - 4x^3 - 12x^2 + 4x^3 - 2x^2 \\ &\quad - 6x - 10x^2 + 5x + 15 \\ &= 8x^4 - 24x^2 - x + 15\end{aligned}$$

$$\begin{aligned}(x - 2)(x^2 + 2x + 4) &= x \cdot x^2 + x \cdot 2x + x \cdot 4 - 2 \cdot x^2 - 2 \cdot 2x - 2 \cdot 4 \\ &= x^3 + 2x^2 + 4x - 2x^2 - 4x - 8 \\ &= x^3 - 8\end{aligned}$$

For convenience, arrange the terms of both polynomials in descending (or ascending) powers of one of the variables.

EXERCISE 12.4

For 1 to 10, multiply as indicated.

1. $(x + 3)(x^2 - 6x + 9)$
2. $(2z^2 - z - 3)(4z^2 + 2z - 5)$
3. $(ax + b)(cx + d)$
4. $(5x - 4)(2x - 3)$
5. $(3x^2 + 2x - 7)(x - 8)$
6. $(3m^2 - 4n^2)(2m^2 - 3n^2)$
7. $(a - b)(a^2 + ab + b^2)$
8. $(z^3 - z^2 + z - 1)(z^2 - z + 1)$
9. $(x^3 - x^2 - x)(x^2 + 2x - 3)$
10. $(2y^2 - 5y + 3)(-3y - 4)$

For 11 to 15, determine the answer in terms of the given variable or variables.

11. Find the product of $a + b - \sqrt{2}c$ and $a + b + \sqrt{2}c$.
12. Multiply $z^4 + 2z^3 - 3z^2 + 7z + 5$ by $3z - 1$.
13. Multiply $3m^2 + 5n^2$ by $2m^2 - 7n^2$.
14. Find the product of $x - y - z$ and $x + y - z$
15. Find the volume of a rectangular box that has dimensions of $x + 4$, $2x - 1$, and $5x$. Hint: $V = lwh$.

Special Products

Memorizing special products is a winning strategy in algebra!

Here is a list of **special products** that you need to know for algebra.

Perfect Squares.

$$(x + y)^2 = x^2 + 2xy + y^2$$

$$(x - y)^2 = x^2 - 2xy + y^2$$

Difference of Two Squares.

$$(x + y)(x - y) = x^2 - y^2$$

Perfect Cubes.

$$(x + y)^3 = x^3 + 3x^2y + 3xy^2 + y^3$$

$$(x - y)^3 = x^3 - 3x^2y + 3xy^2 - y^3$$

Sum of Two Cubes.

$$(x + y)(x^2 - xy + y^2) = x^3 + y^3$$

Difference of Two Cubes.

$$(x - y)(x^2 + xy + y^2) = x^3 - y^3$$

EXERCISE 12.5

For 1 to 10, multiply as indicated.

1. $(a + 6)(a - 6)$

2. $(z + 3)^2$

3. $(x - 2)^2$

4. $(x + 2)(x^2 - 2x + 4)$

5. $(x - 2)(x^2 + 2x + 4)$

6. $(x + 3)^3$

7. $(a - b)(a^2 + ab + b^2)$

8. $(z - 2)^3$

9. $(2m - 1)^2$

10. $(a - 1)(a^2 + a + 1)$

For 11 to 15, determine the answer in terms of the given variable or variables.

11. Find the product of $(x - \sqrt{3})$ and $(x + \sqrt{3})$.

12. Find the square of $(2a + 3)$.

13. Find the cube of $m + 1$.

14. Multiply $4x^2 + 2x + 1$ by $2x - 1$.

15. Find the area of a square whose side measures $x + 4$, in yards. Hint: $A = s^2$.

CHAPTER 13

Simplifying Polynomial Expressions

Removing Parentheses by Addition or Subtraction

EXAMPLE

$5x + (2x - 3) = 5x + 2x - 3 = 7x - 3$

$3x^2 - 2x - (2x^2 - 4x + 5) = 3x^2 - 2x - 2x^2 + 4x - 5 = x^2 + 2x - 5$

$$\begin{aligned} 3z - [4y + 3 - (-2z + 9y - 1) + z] &= 3z - [4y + 3 + 2z - 9y + 1 + z] \\ &= 3z - [-5y + 3z + 4] \\ &= 3z + 5y - 3z - 4 \\ &= 5y - 4 \end{aligned}$$

Start with the innermost parentheses.

EXERCISE 13.1

Remove parentheses, and then, if possible, combine like terms.

1. $6x + (5y + 10)$
2. $9 + (-5y + 4)$
3. $-2a^2 - (-a^2 + 4a)$
4. $2z^2 - 4 - (-5 + 7z^2)$
5. $8a^4 - (2a^4 - 5) + (2a^2 - 1) - 6$
6. $4x + [3 - (2x - 5)]$
7. $4x - [3 - (2x - 5)]$
8. $3x - 4y + [2x - (3x - 4y)] - (5x - 7y)$
9. $m^2 - [m + (2m^2 - 1)] + 3m + [2m - (m^2 - 1)] - 3$
10. $a - (a^2 - (a + 1)) + a^2$

Removing Parentheses by Multiplication or Raising to a Power

Use the order of operations to remove parentheses by multiplication or raising to a power.

EXAMPLE

- $4x + 5(x - 2) = 4x + 5x - 10 = 9x - 10$
- $2z - 5z(3z - 2) + 10z^2 = 2z - 15z^2 + 10z + 10z^2 = -5z^2 + 12z$
- $(m + 1)^2 + (m - 8)(m + 2) = m^2 + 2m + 1 + m^2 - 6m - 16 = 2m^2 - 4m - 15$
- $x^2 - x - 1(5x - 12) = x^2 - x - 5x + 12 = x^2 - 6x + 12$

EXERCISE 13.2

Remove parentheses, and then, if possible, combine like terms.

1. $6x + 2(5x - 3)$
2. $9 - 1(-5z + 4) + 3z$
3. $3a^2 - (-a^2 + 4a + 1) + 2a - a(a + 3) - 8$
4. $8z^2 - 4 - 2(-5 + 7z^2)$
5. $(x + 2)(x - 2) + (x + 2)^2$
6. $(3m - 5)(4m - 1) + (m - 3)(m + 4)$
7. $4x + 2[3x - 2(2x - 5)]$
8. $3x^2 - 4y^2 - x[2x - (3x - 4)] - y(5 - 7y)$
9. $3x(x^2 - 9) - (x - 3)(x^2 + 3x + 9)$
10. $a^2 - a - 4(2a - (a + 1) + 1) + 3a(a - 1)$

CHAPTER 14

Dividing Polynomials

Dividing a Polynomial by a Monomial

To divide a polynomial by a monomial, divide each term of the polynomial by the monomial.

EXAMPLE

$$\frac{16x^3 - 28x^2}{-4x} = \frac{16x^3}{-4x} \underset{\text{Insert}}{+} \frac{\overset{\text{Keep with 28}}{-28x^2}}{-4x} = -4x^2 + 7x$$

$$\frac{-12x^4 + 6x^2}{-3x} = \frac{-12x^4}{-3x} + \frac{6x^2}{-3x} = 4x^3 - 2x$$

$$\frac{6x^4 + 1}{2x^4} = \frac{6x^4}{2x^4} + \frac{1}{2x^4} = 3 + \frac{1}{2x^4}$$

To avoid sign errors when dividing a polynomial by a monomial, *keep a − symbol with the number that follows it*. You likely will need to properly insert a + symbol when you do this.

EXERCISE 14.1

For 1 to 10, multiply as indicated.

1. $\dfrac{4x^4y - 8x^3y^3 + 16xy^4}{4xy}$

2. $\dfrac{16x^5y^2}{16x^5y^2}$

3. $\dfrac{15x^5 - 30x^2}{-5x}$

4. $\dfrac{-14x^4 + 21x^2}{-7x^2}$

5. $\dfrac{25x^4y^2}{-5x}$

6. $\dfrac{6x^5y^2 - 8x^3y^3 + 10xy^6}{2xy^2}$

7. $\dfrac{-10x^4y^4z^4 - 20x^2y^5z^2}{10x^2y^3z}$

8. $\dfrac{-1.8x^5}{0.3x}$

9. $\dfrac{-18x^5 + 5}{3x^5}$

10. $\dfrac{7a^6b^3 - 14a^5b^2 - 42a^4b^2 + 7a^3b^2}{7a^3b^2}$

For 11 to 15, determine the answer in terms of the given variable or variables.

11. Find the quotient of $14x^5y$ and $-xy$.

12. Divide $9ab^4$ by $-3a^2b$.

13. Find the quotient of $3x^2 - 15x$ and $3x$.

14. The area of a rectangle is $24m^2$ and its length is $6m$. Find the rectangle's width. Hint: $A = lw$.

15. If a vehicle travels $25x$ miles in 2 hours, what is the vehicle's average speed, in miles per hour?

Dividing a Polynomial by a Polynomial

You can use long division to divide two polynomials when the divisor is not a monomial. The procedure is very similar to the long division algorithm of arithmetic.

Step 1. Arrange the terms of both the dividend and the divisor in descending powers of the variable.

Step 2. Divide the first term of the dividend by the first term of the divisor, and record the result as the first term of the quotient.

Step 3. Multiply all terms of the divisor by the first term of the quotient, and enter the product under the dividend.

Step 4. Subtract the product in Step 3 from the entire dividend.

Step 5. Treat the difference in Step 4 as a new dividend; and repeat Steps 2, 3, and 4 until the remainder is no longer divisible.

EXAMPLE

Divide $4x^3 + 8x - 6x^2 + 1$ by $2x - 1$.

Step 1.

$$2x - 1 \overline{\big) 4x^3 - 6x^2 + 8x + 1}$$

Step 2.

$$\begin{array}{r l} & 2x^2 \\ 2x - 1 \big) & \overline{4x^3 - 6x^2 + 8x + 1} \end{array}$$

Step 3.

$$\begin{array}{r l} & 2x^2 \\ = 2x - 1 \big) & \overline{4x^3 - 6x^2 + 8x + 1} \\ & 4x^3 - 2x^2 \end{array}$$

Step 4.

$$\begin{array}{r l} & 2x^2 \\ = 2x - 1 \big) & \overline{4x^3 - 6x^2 + 8x + 1} \\ & \underline{4x^3 - 2x^2} \\ & \quad - 4x^2 + 8x + 1 \end{array}$$

Repeat Steps 2 to 4.

$$\begin{array}{r l} & 2x^2 - 2x \\ = 2x - 1 \big) & \overline{4x^3 - 6x^2 + 8x + 1} \\ & \underline{4x^3 - 2x^2} \\ & \quad - 4x^2 + 8x + 1 \\ & \quad \underline{-4x^2 + 2x} \\ & \qquad\qquad 6x + 1 \end{array}$$

Repeat Steps 2 to 4.

$$\begin{array}{r l} & 2x^2 - 2x + 3 \\ = 2x - 1 \big) & \overline{4x^3 - 6x^2 + 8x + 1} \\ & \underline{4x^3 - 2x^2} \\ & \quad - 4x^2 + 8x + 1 \\ & \quad \underline{- 4x^2 + 2x} \\ & \qquad\qquad 6x + 1 \\ & \qquad\qquad \underline{6x - 3} \\ & \qquad\qquad\qquad 4 \end{array}$$

Answer

$$2x^2 - 2x + 3 + \frac{4}{2x - 1}$$

EXAMPLE

Insert zeros as placeholders for missing powers of x.

Divide $x^3 - 8$ by $x - 2$.

Step 1.

$$= x - 2\overline{)x^3 + 0 + 0 - 8}$$

Step 2.

$$\begin{array}{r} x^2 \\ = x - 2\overline{)x^3 + 0 + 0 - 8} \end{array}$$

Step 3.

$$\begin{array}{rl} & x^2 \\ = x - 2 & \overline{)x^3 + 0 + 0 - 8} \\ & x^3 - 2x^2 \end{array}$$

Step 4.

$$\begin{array}{rl} & x^2 \\ = x - 2 & \overline{)x^3 + 0 + 0 - 8} \\ & \underline{x^3 - 2x^2} \\ & \phantom{)x^3 -{}} 2x^2 + 0 - 8 \end{array}$$

Repeat Steps 2 to 4.

$$\begin{array}{rl} & x^2 + 2x \\ = x - 2 & \overline{)x^3 + 0 + 0 - 8} \\ & \underline{x^3 - 2x^2} \\ & \phantom{)x^3 -{}} 2x^2 + 0 - 8 \\ & \phantom{)x^3 -{}} \underline{2x^2 - 4x} \\ & \phantom{)x^3 - 2x^2 +{}} 4x - 8 \end{array}$$

Repeat Steps 2 to 4.

$$\begin{array}{rl} & x^2 + 2x + 4 \\ = x - 2 & \overline{)x^3 + 0 + 0 - 8} \\ & \underline{x^3 - 2x^2} \\ & \phantom{)x^3 -{}} 2x^2 + 0 - 8 \\ & \phantom{)x^3 -{}} \underline{2x^2 - 4x} \\ & \phantom{)x^3 - 2x^2 +{}} 4x - 8 \\ & \phantom{)x^3 - 2x^2 +{}} \underline{4x - 8} \\ & \phantom{)x^3 - 2x^2 + 4x - {}} 0 \end{array}$$

Answer

$x^2 + 2x + 4$

EXERCISE 14.2

Answer the following questions.

1. Divide $x^2 + x - 4$ by $(x + 3)$.
2. Find the quotient of $x^2 - 9x + 20$ and $x - 5$.
3. Divide $2x^3 - 13x + x^2 + 6$ by $x - 4$.
4. Divide $x^6 - 64$ by $x^3 - 8$.
5. If the area of a triangle is $x^2 + x - 12$ and its base is $x - 3$, then express its height in terms of x.

CHAPTER 15

Factoring Polynomials

Factoring Out the Greatest Common Factor

The **greatest common factor (GCF)** of a polynomial is the greatest factor that is common to the polynomial's terms.

EXAMPLE

- $10x^2 - 5x = 5x(2x - 1)$
- $-a - b = -1(a + b)$
- $-12x^3 + 8x^2 - 16x = -4x(3x^2 - 2x + 4)$
- $a(x + y) + b(x + y) = (x + y)(a + b)$
- $z^3 + z^2 + 2z + 2 = z^2(z + 1) + 2(z + 1) = (z + 1)(z^2 + 2)$

EXERCISE 15.1

For 1 to 10, factor out the greatest common factor using the GCF with a positive coefficient.

1. $4x + 4y$
2. $3x + 6$
3. $12x^8y^3 - 8x^6y^7z^2$
4. $15x^2 - 3x$
5. $x^3y - xy + y$
6. $\frac{1}{2}ax - \frac{1}{2}ay$
7. $x(w - z) - y(w - z)$
8. $1.5a^2b + 4.5ab + 7.5ab^2$
9. $mx + my + 5x + 5y$
10. $xy + xy^2 + xy^3 + xy^4$

For 11 to 15, factor out the greatest common factor using the GCF with a negative coefficient.

11. $-5x - 5y$
12. $-4x + 8$
13. $-24x^8y^3 - 8x^6y^7z^2$
14. $15x^2 - 3x$
15. $ar - rt - r$

Factoring the Difference of Two Squares

Use your knowledge of special products to factor the difference of two squares.

EXAMPLE

- $x^2 - 9 = (x + 3)(x - 3)$
- $4x^2 - 25 = (2x + 5)(2x - 5)$
- $\frac{1}{4}x^2 - 49y^2 = \left(\frac{1}{2}x + 7y\right)\left(\frac{1}{2}x - 7y\right)$
- $x^2 - 0.25y^2 = (x + 0.5y)(x - 0.5y)$
- $x^2 - 2 = \left(x + \sqrt{2}\right)\left(x - \sqrt{2}\right)$

Polynomials of the form $x^2 + a^2$, where a is a real number, are not factorable over the real numbers.

EXERCISE 15.2

Factor the difference of two squares.

1. $16x^2 - 36$
2. $x^2 - y^2$
3. $36x^2 - 49$
4. $x^2 - 1$
5. $\frac{1}{4}x^2 - 25$
6. $z^2 - 0.36$
7. $100x^2y^2 - 81z^2$
8. $a^2 - \frac{1}{49}$
9. $x^2 - 3$
10. $z^2 - 5$
11. $121 - 25c^2$
12. $x^4 - y^2$
13. $x^2y^2 - z^2$
14. $x^4y^2 - 1$
15. $0.64 - x^2$

Factoring Perfect Trinomial Squares

Use your knowledge of special products to factor perfect trinomial squares.

EXAMPLE

- $x^2 - 10x + 25 = (x - 5)^2$
- $x^2 - 6x + 9 = (x - 3)^2$
- $4x^2 + 4x + 1 = (2x + 1)^2$

EXERCISE 15.3

Factor the perfect square.

1. $x^2 - 4x + 4$
2. $x^2 + 10xy + 25y^2$
3. $36x^2 + 12x + 1$
4. $4x^2 - 12x + 9$
5. $49a^2 + 56a + 16$
6. $x^2 + x + \frac{1}{4}$
7. $100 + 140x - 49x^2$
8. $36x^2 + 60x + 25$
9. $81 - 36x + 4x^2$
10. $1 - 2x + x^2$

Factoring the Sum and the Difference of Two Cubes

Use your knowledge of special products to factor the sum and difference of two cubes.

EXAMPLE

- $x^3 + 27 = (x + 3)(x^2 - 3x + 9)$
- $x^3 - 8 = (x - 2)(x^2 + 2x + 4)$
- $27x^3 - 8y^3 = (3x - 2y)(9x^2 + 6xy + 4y^2)$

EXERCISE 15.4

Factor the sum or difference of two cubes.

1. $x^3 + 125$
2. $y^3 - 27$
3. $64a^3 + 1$
4. $8z^3 - 125$
5. $125a^3 + 27$
6. $8x^3 - 125$
7. $64y^3 - 125$
8. $216x^3 + 1,000$
9. $8x^3 + 125y^3z^6$
10. $125a^6 - 27$

Factoring General Trinomials

Use your knowledge of FOIL to factor general trinomials.

EXAMPLE

- $x^2 + 7x + 10 = (x + 2)(x + 5)$
- $x^2 - 7x + 10 = (x - 2)(x - 5)$
- $x^2 - 3x - 10 = (x + 2)(x - 5)$
- $2x^2 + x - 6 = (x + 2)(2x - 3)$

EXERCISE 15.5

Factor the trinomial.

1. $x^2 + 5x + 6$
2. $x^2 + 5x - 6$
3. $y^2 + 9y - 10$
4. $b^2 + 7b - 98$
5. $z^2 - 5z - 14$
6. $2x^2 + 5x - 3$
7. $6x^2 - x - 1$
8. $9y^2 + 9y - 4$
9. $49b^2 - 21b - 10$
10. $16z^2 - 16z - 5$

CHAPTER 16

Fundamental Concepts of Rational Expressions

Definition of a Rational Expression

A **rational expression** is an algebraic expression that can be written as the ratio of two polynomials. Because division by 0 is undefined, you must exclude all real values for the variable (or variables) that would make the denominator polynomial evaluate to 0. Be mindful of this restriction.

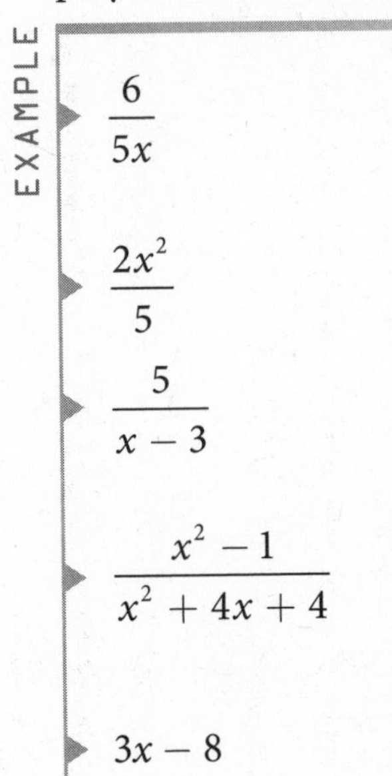

EXAMPLE

- $\dfrac{6}{5x}$ → A rational expression with the restriction $x \neq 0$.
- $\dfrac{2x^2}{5}$ → A rational expression with no restrictions on the variable x.
- $\dfrac{5}{x-3}$ → A rational expression with the restriction $x \neq 3$.
- $\dfrac{x^2-1}{x^2+4x+4}$ → A rational expression with the restriction $x \neq -2$.
- $3x - 8$ → A rational expression with no restrictions on the variable x.

Note: $3x - 8 = \dfrac{3x-8}{1}$

EXERCISE 16.1

State the restrictions, if any, for the following rational expressions.

1. $\frac{3x-1}{x+4}$
2. $\frac{5}{x^2+3}$
3. $\frac{7x-4}{x^2-2x+1}$
4. $\frac{8x}{x^2+5x+6}$
5. $\frac{9}{x^2}$
6. x^2-4
7. $\frac{3}{x^2-4}$
8. $\frac{9}{x^2+9}$
9. $\frac{x-5}{x^2-9}$
10. $\frac{3x}{2x+5}$

Reducing Rational Expressions

Before applying the Fundamental Principle of Rational Expressions, *always* make sure that you have only *factored* polynomials in the numerator and denominator.

Fundamental Principle of Rational Expressions. If P, Q, and R are polynomials, then $\frac{PR}{QR} = \frac{RP}{RQ} = \frac{P}{Q}$, provided neither Q nor R has a zero value. Hereafter, for convenience, you can assume restricted values are excluded as you work through the problems in this chapter.

The fundamental principle allows you to **reduce rational expressions to lowest terms** by dividing the numerator and denominator by the GCF.

EXAMPLE

$$\frac{15x^5y^3z}{30x^5y^3} = \frac{15x^5y^3 \cdot z}{15x^5y^3 \cdot 2} = \frac{\cancel{15x^5y^3} \cdot z}{\cancel{15x^5y^3} \cdot 2} = \frac{z}{2}$$

$$\frac{6x}{2x} = \frac{2x \cdot 3}{2x \cdot 1} = \frac{\cancel{2x} \cdot 3}{\cancel{2x} \cdot 1} = \frac{3}{1} = 3$$

$$\frac{x-3}{3-x} = \frac{x-3}{-1(-3+x)} = \frac{x-3}{-1(x-3)} = \frac{(x-3)}{-1(x-3)} = \frac{1\cancel{(x-3)}}{-1\cancel{(x-3)}} = -1$$

$$\frac{2x+6}{x^2+5x+6} = \frac{2(x+3)}{(x+2)(x+3)} = \frac{2\cancel{(x+3)}}{(x+2)\cancel{(x+3)}} = \frac{2}{x+2}$$

$\frac{2}{x+2} \neq \frac{1}{x+1}$; 2 is a factor of the numerator, but it is a *term* of the denominator. It is a mistake to divide out a term.

$$\frac{x^2-1}{x^2+2x+1}=\frac{(x+1)(x-1)}{(x+1)(x+1)}=\frac{\cancel{(x+1)}(x-1)}{\cancel{(x+1)}(x+1)}=\frac{x-1}{x+1}$$

$$\frac{x(a-b)+2(a-b)}{(a-b)}=\frac{(a-b)(x+2)}{1(a-b)}=\frac{\cancel{(a-b)}(x+2)}{1\cancel{(a-b)}}=\frac{(x+2)}{1}=x+2$$

EXERCISE 16.2

Reduce to the lowest terms.

1. $\frac{-8x^2y}{16xy^2}$

2. $\frac{24a^3x^2}{32ay}$

3. $\frac{18x^4}{30x}$

4. $\frac{10x^2z^3}{xz^4}$

5. $\frac{5x^2(x+3)}{15x(x+3)}$

6. $\frac{12xy^2}{6x^4(x-y)}$

7. $\frac{xy-x^2y^2}{xz-x^2yz}$

8. $\frac{\frac{4}{3}\pi r^3}{4\pi r^2}$

9. $\frac{4x(y-2z)}{2x^4(y-2z)}$

10. $\frac{10a^2b^5}{5b^2(a+b)}$

11. $\frac{x^2-16}{x^2-8x+16}$

12. $\frac{z^2+4z-5}{z^2+8z+15}$

13. $\frac{2y^2+4y-30}{3y^2+21y+30}$

14. $\frac{3xy^3-27xy}{6xy^2+6xy-72x}$

15. $\frac{a^2-12a+36}{a^2-3a-18}$

16. $\frac{4z^2+16z+16}{6z^2+18z+12}$

17. $\frac{x^3-xy^2}{xy(x^2-2xy+y^2)}$

18. $\frac{3t+15}{t^2-25}$

19. $\frac{2x-10}{x^2-10x+25}$

20. $\frac{(x-2)}{(x^3-8)}$

Building Up the Denominator of a Rational Expression

To build up the denominator of a fraction, multiply the numerator and denominator by the same nonzero expression to obtain an equivalent fraction that has the desired denominator.

EXAMPLE

$$\frac{5}{6x} = \frac{?}{18xy}$$

$$\frac{5}{6x} = \frac{5 \cdot 3y}{6x \cdot 3y} = \frac{15y}{18xy}$$

EXAMPLE

$$\frac{x-1}{x+2} = \frac{?}{x^2 - x - 6}$$

$$\frac{(x-1)}{(x+2)} = \frac{?}{(x+2)(x-3)}$$

$$\frac{(x-1)(x-3)}{(x+2)(x-3)} = \frac{x^2 - 4x + 3}{x^2 - x - 6}$$

EXAMPLE

$$\frac{x}{5-x} = \frac{?}{x^2 - 25}$$

$$\frac{x}{5-x} = \frac{?}{(x+5)(x-5)}$$

$$\frac{(-1)x}{(-1)(5-x)} = \frac{?}{(x+5)(x-5)}$$

$$\frac{-x}{(x-5)} = \frac{-x(x+5)}{(x+5)(x-5)} = \frac{-x^2 - 5x}{x^2 - 25}$$

EXERCISE 16.3

Convert the given rational expression into an equivalent one with the indicated denominator.

1. $\frac{x}{2y} = \frac{?}{16xy^2}$

2. $\frac{2y}{3x^2} = \frac{?}{12x^5y}$

3. $\frac{3x^3}{5} = \frac{?}{30x}$

4. $\frac{10xz}{1} = \frac{?}{xz^4}$

5. $\frac{1}{x^3} = \frac{?}{x^8}$

6. $\frac{x}{3xy} = \frac{?}{18xy}$

7. $\frac{1}{x+3} = \frac{?}{(x+3)^2}$

8. $\frac{x+3}{x-3} = \frac{?}{x^2-9}$

9. $\frac{3x(x+2)}{(x-2)} = \frac{?}{x^2-6x+8}$

10. $\frac{2a^2}{b^2} = \frac{?}{5ab^2+5b^3}$

11. $\frac{x+4}{x-4} = \frac{?}{x^2-8x+16}$

12. $\frac{z+2}{z+5} = \frac{?}{z^2+8z+15}$

13. $\frac{2y-6}{3y+6} = \frac{?}{3y^2+21y+30}$

14. $2x^2y = \frac{?}{2x}$

15. $\frac{x+y}{2xy^2} = \frac{?}{6x^2y^2}$

16. $\frac{5}{x+3} = \frac{?}{x^2+x-6}$

17. $\frac{2a}{a+5} = \frac{?}{a^2+2a-15}$

18. $\frac{4c}{c-2} = \frac{?}{(c-2)^2}$

19. $\frac{x}{3(x+4)} = \frac{?}{6(x+4)(4x+1)}$

20. $\frac{5x}{(x+4)(x+3)} = \frac{?}{(x+4)(x+3)(x-1)}$

CHAPTER 17

Multiplying and Dividing Rational Expressions

Multiplying Rational Expressions

To multiply rational expressions, factor all numerators and denominators completely, as needed; divide numerators and denominators by their common factors; and then multiply the remaining numerator factors to get the numerator of the product and multiply the remaining denominator factors to get the denominator of the product.

When you are multiplying rational expressions, if a numerator or denominator does not factor, enclose it in parentheses. Omitting the parentheses can lead to a mistake.

EXAMPLE

$$\frac{5x^2}{2y^3}\cdot\frac{y^4}{10x^5}=\frac{5x^2}{2y^3}\cdot\frac{y^4}{10x^5}=\frac{y}{4x^3}$$

$$3x^2y\cdot\frac{3x-4y}{6x}=\frac{3x^2y}{1}\cdot\frac{(3x-4y)}{6x}=\frac{3x^2y-4xy^2}{2}$$

$$\frac{x^2-9}{x^2+6x+9}\cdot\frac{2x+6}{x^2-x-6}=\frac{(x+3)(x-3)}{(x+3)(x+3)}\cdot\frac{2(x+3)}{(x-3)(x+2)}=\frac{2}{x+2}$$

$$\frac{x^2-2x+1}{x^2-4}\cdot\frac{3x-6}{5x-5}=\frac{(x-1)(x-1)}{(x+2)(x-2)}\cdot\frac{3(x-2)}{5(x-1)}=\frac{3(x-1)}{5(x+2)}$$

When you multiply rational expressions, you may find it convenient to leave your answer in factored form. Always double-check to make sure it is in completely reduced form.

EXERCISE 17.1

Multiply as indicated.

1. $\frac{2a}{3b} \cdot \frac{6b}{a}$

2. $\frac{24x^2}{48y^2} \cdot y^2$

3. $\frac{4x^4}{5} \cdot \frac{10}{x^3}$

4. $\frac{27x^2y^3}{8z^3} \cdot \frac{16x^2}{9xy}$

5. $\frac{2a^2x^4}{9ab^2} \cdot \frac{3x^2b^3}{8ax^3} \cdot \frac{6bx^2}{7a^5b}$

6. $\frac{a+b}{a-b} \cdot \frac{a-b}{a+b}$

7. $\frac{x^2-5x-6}{x^2+6x+8} \cdot \frac{x^2+x-2}{x^2-2x-3}$

8. $\frac{x+y}{3xy} \cdot \frac{6xy}{(x+y)^2}$

9. $\frac{a+6}{a-3} \cdot (a-3)$

10. $\frac{(x-2)^5}{4x} \cdot \frac{12x^3}{(x-2)^3}$

11. $\frac{(x+y)(y+2)}{(x+2)} \cdot \frac{(x+2)(x+2y)}{2(y+2)}$

12. $\frac{(x-2)(y+2)}{(x+2)(y-2)} \cdot \frac{(x+2)(y-2)}{(x-2)(y-2)}$

13. $\frac{4x^2+10}{x-3} \cdot \frac{x^2-9}{6x^2+15}$

14. $\frac{4x-4}{x^2-y^2} \cdot \frac{x^3y^2-x^2y^3}{2x-2}$

15. $\frac{a^2-12a+36}{3a^3} \cdot \frac{6a}{a^2-3a-18}$

16. $\frac{3}{6z^2+18z+12} \cdot \frac{4z^2+16z+16}{4}$

17. $\frac{1}{(x^2-2xy+y^2)} \cdot \frac{x^3-xy^2}{xy}$

18. $\frac{t-5}{3t^2-75} \cdot \frac{6t+30}{2t}$

19. $\frac{x^2-7x+12}{x^2-x-6} \cdot \frac{x^2+7x+10}{x^2+x-20}$

20. $\frac{x^2-13x+42}{x^2+2x} \cdot \frac{x^2+x-2}{2x^2-14x}$

Dividing Rational Expressions

To divide rational expressions, multiply the first rational expression (the dividend) by the reciprocal of the second rational expression (the divisor).

EXAMPLE

$$\frac{2xy^2}{3z^3} \div \frac{4x^3y}{9z^2} = \frac{2xy^2}{3z^3} \cdot \frac{9z^2}{4x^3y} = \frac{3y}{2x^2z}$$

$$\frac{x^2-2x+1}{x^2-x-6} \div \frac{x^2-3x+2}{x^2-4} = \frac{x^2-2x+1}{x^2-x-6} \cdot \frac{x^2-4}{x^2-3x+2}$$

$$= \frac{(x-1)(x-1)}{(x-3)(x+2)} \cdot \frac{(x+2)(x-2)}{(x-1)(x-2)} = \frac{x-1}{x-3}$$

EXERCISE 17.2

Divide as indicated.

1. $\frac{3}{x} \div \frac{5}{x}$
2. $\frac{4xy}{3y} \div \frac{2x}{y}$
3. $\frac{3x^2y}{8z^2} \div \frac{6xy^3}{z}$
4. $5x^2y \div \frac{10xy^2}{3}$
5. $\frac{2m^3n}{11} \div \frac{4mn^3}{33}$
6. $\frac{18}{x} \div 6$
7. $\frac{2x^4y^2}{5z} \div 8xy^2$
8. $\frac{4}{3}\pi r^3 \div 4\pi r^2$
9. $4xyz \div \frac{2x^2y}{3z^2}$
10. $\frac{9a^2b^4}{10c} \div 27abc$
11. $\frac{4x-4y}{5x+5y} \div \frac{4}{25}$
12. $\frac{4x-4}{x^2-16} \div \frac{x-1}{x-4}$
13. $\frac{x^2-y^2}{15(x+y)^2} \div \frac{x-y}{5x+5y}$
14. $\frac{x^2+4x-12}{x^2+9x+18} \div \frac{3x+12}{6x+6}$
15. $\frac{z^2+14z+49}{z^2+2z-35} \div \frac{z^2+9z+14}{z^2-3z-10}$
16. $\frac{x^3y+2x^2y^2+xy^3}{x^4-y^4} \div \frac{x^2+xy}{x^2+y^2}$
17. $\frac{a+b}{a^2-ab} \div \frac{1}{a^2-b^2}$
18. $\frac{x^2+x-20}{4x^3} \div \frac{x^2-16}{6x^2}$
19. $\frac{7m^2n^2}{8} \div \frac{21mn^4}{16}$
20. $\frac{x^2+6x+9}{x^2+2x-3} \div \frac{x^2-9}{x^2-x-6}$

CHAPTER 18

Adding and Subtracting Rational Expressions

Adding and Subtracting Rational Expressions, Like Denominators

To add (or subtract) rational expressions that have the same denominators, place the sum (or difference) of the numerators over the common denominator. Simplify and reduce to lowest terms, as needed.

> When subtracting algebraic fractions, you must enclose the numerator of the second fraction in parentheses because you want to subtract the *entire numerator*, not just the first term.

EXAMPLE

$$\frac{x+2}{x-3}+\frac{2x-11}{x-3}=\frac{3x-9}{x-3}=\frac{3\cancel{(x-3)}}{\cancel{(x-3)}}=\frac{3}{1}=3$$

$$\frac{5x^2}{4(x+1)}-\frac{4x^2+1}{4(x+1)}=\frac{(5x^2)-(4x^2+1)}{4(x+1)}=\frac{x^2-1}{4(x+1)}=\frac{\cancel{(x+1)}(x-1)}{4\cancel{(x+1)}}$$

$$=\frac{(x-1)}{4}$$

EXERCISE 18.1

Add or subtract as indicated.

1. $\frac{5x}{9} + \frac{2x}{9}$

2. $\frac{2a}{5y} + \frac{3a}{5y}$

3. $\frac{x}{x+5} + \frac{4x}{x+5}$

4. $\frac{x}{(x-2)(x+2)} + \frac{2}{(x-2)(x+2)}$

5. $\frac{x^2}{x^2-25} + \frac{10x+25}{x^2-25}$

6. $\frac{x}{3x+1} + \frac{2x}{3x+1} + \frac{1}{3x+1}$

7. $\frac{2x+2}{5} + \frac{2x+1}{5} + \frac{x+3}{5}$

8. $\frac{x^2+x-1}{x^2+2x+1} + \frac{x^2+x+1}{x^2+2x+1}$

9. $\frac{5x}{10a} + \frac{x}{10a} + \frac{4x}{10a}$

10. $\frac{x^2+5x-13}{x^2+3x-10} + \frac{x^2+x-7}{x^2+3x-10}$

11. $\frac{x}{x^2-16} - \frac{4}{x^2-16}$

12. $\frac{3x}{(x-3)^2} - \frac{2x}{(x-3)^2}$

13. $\frac{x^2-xy}{x^2-y^2} - \frac{xy-y^2}{x^2-y^2}$

14. $\frac{x^2}{x^2+2x-15} - \frac{25}{x^2+2x-15}$

15. $\frac{z^2}{z^2-9} - \frac{6z-9}{z^2-9}$

16. $\frac{7x+4y}{10} - \frac{2x+3y}{10}$

17. $\frac{x(x-y)}{x^2-y^2} - \frac{y(y-x)}{x^2-y^2}$

18. $\frac{4x^2}{x^2-8x+16} - \frac{64}{x^2-8x+16}$

19. $\frac{7x}{5(x+3)} + \frac{4x}{5(x+3)} - \frac{x}{5(x+3)}$

20. $\frac{x^2+6x+2}{(x+3)(x-2)} - \frac{2x-1}{(x+3)(x-2)}$

Adding and Subtracting Rational Expressions, Unlike Denominators

To add (or subtract) rational expressions that have different denominators: first, factor each denominator completely; next, find the least common denominator, which is the least product that is divisible by each denominator; then, convert each rational expression to one having the common denominator as a denominator; and, finally, add (or subtract) as for like denominators.

EXAMPLE

$$\frac{3x}{x^2-4}+\frac{x}{x-2}=\frac{3x}{(x+2)(x-2)}+\frac{x\cdot(x+2)}{(x-2)\cdot(x+2)}$$
$$=\frac{3x+x^2+2x}{(x+2)(x-2)}$$
$$=\frac{x^2+5x}{(x+2)(x-2)}$$
$$=\frac{x(x+5)}{(x+2)(x-2)}$$

EXAMPLE

$$\frac{2x-1}{x-3}-\frac{x}{2x+2}=\frac{2x-1}{(x-3)}-\frac{x}{2(x+1)}$$
$$=\frac{(2x-1)\cdot 2(x+1)}{(x-3)\cdot 2(x+1)}-\frac{x\cdot(x-3)}{2(x+1)\cdot(x-3)}$$
$$=\frac{4x^2+2x-2}{2(x-3)(x+1)}-\frac{x^2-3x}{2(x-3)(x+1)}$$
$$=\frac{4x^2+2x-2-(x^2-3x)}{2(x-3)(x+1)}$$
$$=\frac{4x^2+2x-2-x^2+3x}{2(x-3)(x+1)}$$
$$=\frac{3x^2+5x-2}{2(x-3)(x+1)}$$
$$=\frac{(3x-1)(x+2)}{2(x-3)(x+1)}$$

EXERCISE 18.2

Add or subtract as indicated.

1. $\dfrac{2x}{5}+\dfrac{3x}{10}$

2. $\dfrac{5m}{12x^2y}-\dfrac{3n}{10xy^2}$

3. $\dfrac{3a+6}{2a}+\dfrac{5b+4}{2b}$

4. $\dfrac{x+y}{2x}-\dfrac{x-y}{3y}+\dfrac{y-z}{z}$

5. $\dfrac{4(x+2)}{x}-\dfrac{2(x+4)}{3x}+\dfrac{x+1}{6x}$

6. $\dfrac{4x^2-5}{3x}+2x$

7. $4x^2-\dfrac{12x^2-3}{5x}$

8. $\dfrac{y(x-3)}{4x}+\dfrac{x^2-y^2}{xy}-\dfrac{x(y-2)}{6y}$

9. $\dfrac{x-4}{4}-\dfrac{x-4}{x}$

10. $\dfrac{4}{x+4}+\dfrac{5}{x-2}$

11. $\dfrac{3}{x+1}-\dfrac{5}{x-2}$

12. $\dfrac{x}{x-y}+\dfrac{x+y}{xy}$

13. $\dfrac{3}{x+3}+\dfrac{2}{x+2}+\dfrac{2}{3}$

14. $2x-\dfrac{3x^2+5}{2x+5}$

15. $a+\dfrac{2a^2}{a-5}-6$

16. $\dfrac{2t}{t+4}-\dfrac{3t-5}{t^2+8t+16}$

17. $\dfrac{6z}{z^2+5z+6}-\dfrac{2z}{z^2+6z+9}$

18. $\dfrac{2x}{x^2-25}+\dfrac{4}{x+5}$

19. $\dfrac{3x-1}{2+4x}-\dfrac{x}{4+2x}$

20. $\dfrac{x-2}{2(x^2-9)}+\dfrac{x+3}{3(x^2-x-6)}$

CHAPTER 19

Simplifying Complex Fractions

A **complex fraction** has fractions in its numerator, denominator, or both.

Writing Complex Fractions as Division Problems to Simplify

One way to simplify a complex fraction is to interpret the fraction bar of the complex fraction as meaning division.

EXAMPLE

$$\frac{\frac{1}{x}+\frac{1}{y}}{\frac{1}{x}-\frac{1}{y}}=\left(\frac{1}{x}+\frac{1}{y}\right)\div\left(\frac{1}{x}-\frac{1}{y}\right)=\left(\frac{y+x}{xy}\right)\div\left(\frac{y-x}{xy}\right)=\frac{(y+x)}{\cancel{xy}}\cdot\frac{\cancel{xy}}{(y-x)}=\frac{(y+x)}{(y-x)}$$

$$\frac{xy^{-2}}{2^{-1}}=\frac{\left(\frac{x}{y^2}\right)}{\left(\frac{1}{2}\right)}=\frac{x}{y^2}\div\frac{1}{2}=\frac{x}{y^2}\cdot\frac{2}{1}=\frac{2x}{y^2}$$

EXERCISE 19.1

Simplify the complex fraction.

1. $\dfrac{\left(\dfrac{3}{2}\right)}{\left(\dfrac{15}{16}\right)}$

2. $\dfrac{\dfrac{(x^2 - y^2)}{8}}{\dfrac{(x - y)}{32}}$

3. $\dfrac{\left(\dfrac{x}{y^2}\right)}{\left(\dfrac{x^2}{y}\right)}$

4. $\dfrac{1}{1 - \dfrac{x}{y}}$

5. $\dfrac{1}{\left(\dfrac{1 - x}{y}\right)}$

6. $\dfrac{a}{a - \dfrac{a}{2}}$

7. $\dfrac{5x^{-1} - \dfrac{3}{4}}{x^{-2} - \dfrac{1}{2}}$

8. $\dfrac{x^{-1} + y^{-3}}{x^{-4} + y^{-2}}$

9. $\dfrac{4x^{-1}y}{\left(\dfrac{z^{-2}}{2}\right)}$

10. $\dfrac{\left(\dfrac{9a^2b^4}{10c}\right)}{27abc}$

Using the LCD to Simplify a Complex Fraction

Another way to simplify a complex fraction is to multiply its numerator and denominator by the LCD of all the fractions in its numerator and denominator.

EXAMPLE

$$\frac{x^{-1} + y^{-1}}{x^{-1} - y^{-1}} = \frac{\dfrac{1}{x} + \dfrac{1}{y}}{\dfrac{1}{x} - \dfrac{1}{y}} = \frac{xy\left(\dfrac{1}{x} + \dfrac{1}{y}\right)}{xy\left(\dfrac{1}{x} - \dfrac{1}{y}\right)} = \frac{xy \cdot \dfrac{1}{x} + xy \cdot \dfrac{1}{y}}{xy \cdot \dfrac{1}{x} - xy \cdot \dfrac{1}{y}} = \frac{y + x}{y - x}$$

EXERCISE 19.2

Simplify the complex fraction.

1. $\dfrac{\left(\dfrac{4x-4y}{5x+5y}\right)}{\left(\dfrac{4}{25}\right)}$

2. $\dfrac{\left(\dfrac{4x-4}{x^2-16}\right)}{\left(\dfrac{x-1}{x-4}\right)}$

3. $\dfrac{\left(\dfrac{x^2-y^2}{15(x+y)^2}\right)}{\left(\dfrac{x-y}{5x+5y}\right)}$

4. $\dfrac{5-\dfrac{3}{x}}{\dfrac{1}{2}-\dfrac{1}{x^2}}$

5. $\dfrac{\dfrac{x}{x-3}-\dfrac{x+4}{x+3}}{\dfrac{x}{x^2-9}+\dfrac{6}{x^2-9}}$

6. $\dfrac{\dfrac{1}{2x}-\dfrac{1}{4}}{\dfrac{3}{x}+\dfrac{1}{6}}$

7. $\dfrac{\left(\dfrac{a+b}{a^2-ab}\right)}{\left(\dfrac{1}{a^2-b^2}\right)}$

8. $\dfrac{\left(\dfrac{x^2+x-20}{4x^3}\right)}{\left(\dfrac{x^2-16}{6x^2}\right)}$

9. $\dfrac{x^{-3}}{y^{-3}-x^{-3}}$

10. $\dfrac{\dfrac{x+2}{x-1}-\dfrac{x-3}{x}}{\dfrac{x+4}{x}+\dfrac{x-2}{x-1}}$

CHAPTER 20

One-Variable Linear Equations and Inequalities

Basic Concepts

An **equation** is a statement of equality between two mathematical expressions. Whatever is on the left side of the equal sign is the *left side* of the equation, and whatever is on the right side of the equal sign is the *right side* of the equation. An equation is **true** when both the left side and the right side have the same value. Equations that contain only numbers are either true or false.

EXAMPLE

- $9 + 7 = 16$ is true.
- $9 + 7 = 20$ is false.

A **one-variable equation** is one that contains only one variable that serves as a placeholder for a number or numbers in the equation. A one-variable equation is an **identity** if it is true for all admissible values of the variable. A one-variable equation is a **conditional equation** if it is true for only some values of the variable, but not for other values. A one-variable equation is a **contradiction** if it is false no matter what value is substituted for the variable.

EXAMPLE

- $4x + 5 = 2x + 2x + 5$ is an identity because any value substituted for x will result in a true statement.
- $2x + 3 = 13$ is a conditional equation because it is true only when x equals 5.
- $x^2 = x^2 + 9$ is a contradiction because it is false no matter what value is substituted for x.

EXERCISE 20.1

State whether the equation is true or false for the given value of the variable.

1. $\frac{3}{4} + x = 1.25,\ x = \frac{1}{2}$
2. $2x = 30, x = 6$
3. $x^2 = x^2 + 9, x = 3$
4. $-9.22 + x = 10.25 - 8.43, x = 7.4$
5. $4x + 5 = 2x + 2x + 5, x = 20$
6. $x^2 = 25, x = -5$
7. $\frac{4x - 3 \cdot 2}{6x - 2^2} = \frac{1}{2}, x = 4$
8. $4x + 5 = 2x + 2x + 5, x = -10$
9. $2x = 30, x = 15$
10. $\frac{x - 3}{2} = \frac{2x + 4}{5}, x = 0$
11. $x^2 = x^2 + 9, x = 10$
12. $3x - 2 = 7 - 2x, x = 1.8$
13. $\frac{x - 3}{2} = \frac{2x + 4}{5}, x = 23$
14. $-3x - 7 = 14, x = 7$
15. $5x + 9 = 3x - 1, x = -5$
16. $4(x - 6) = 40, x = 16$
17. $x^2 = -4, x = -2$
18. $x^2 + x = 6, x = -3$
19. $x^2 + x = 6, x = 2$
20. $x^2 = 7,\ x = -\sqrt{7}$

Solving One-Variable Linear Equations

A **one-variable linear equation** in x is one that can be written in the form $ax + b = c, a \neq 0$ where a, b, and c are constants.

EXAMPLE AND NONEXAMPLE

- $-3x - 7 = 14$ → A one-variable linear equation
- $2x + 3 = 13$ → A one-variable linear equation
- $\frac{3}{4} + x = 1.25$ → A one-variable linear equation
- $3x - 2 = 7 - 2x$ → A one-variable linear equation
- $x^2 + x = 6$ → Not a one-variable linear equation
- $x^2 = 7$ → Not a one-variable linear equation

Observe that a one-variable linear equation contains only constants and a single variable raised to the first power.

To **solve** a one-variable linear equation means to find a numerical value for the variable that makes the equation true. The goal in solving a one-variable linear equation is to get the variable by itself on only one side of the equation and with a coefficient of 1 (usually understood).

Use the properties of real numbers and simple algebraic tools to solve equations. An equation is like a balance scale. To keep the equation in balance, when you do something to one side of the equation, you must do the same thing to the other side of the equation. As you solve the equation, undo what has been done to the variable until you get an expression like this: variable = solution.

Tools for Solving One-Variable Linear Equations

- Add the same number to both sides.
- Subtract the same number from both sides.
- Multiply both sides by the same nonzero number.
- Divide both sides by the same nonzero number.

When solving an equation, remember that you must never multiply or divide both sides by zero.

Application of one or more of these tools will yield an equation that is equivalent to the original equation. Equations that have the same solution are **equivalent equations**. Decide which operation to do based on what has been done to the variable. As you proceed, exploit the fact that addition and subtraction undo each other; and, similarly, multiplication and division undo one another.

EXAMPLE

Solve $-3x - 7 = 14$.

$$-3x - 7 = 14$$
$$-3x - 7 + 7 = 14 + 7$$
$$-3x = 21$$
$$\frac{-3x}{-3} = \frac{21}{-3}$$
$$x = -7$$

EXAMPLE

Solve $\frac{3}{5}x = 72$.

$$\frac{3}{5}x = 72$$
$$\frac{5}{3} \cdot \frac{3}{5}x = \frac{5}{3} \cdot \frac{72}{1}$$
$$x = 120$$

EXAMPLE

If the variable appears on both sides of the equation, add or subtract a variable expression to both sides, so that the variable appears on only one side of the equation.

Solve $5x + 9 = 3x - 1$.

$$5x + 9 = 3x - 1$$
$$5x + 9 - 3x = 3x - 1 - 3x$$
$$2x + 9 = -1$$
$$2x + 9 - 9 = -1 - 9$$
$$2x = -10$$
$$\frac{2x}{2} = \frac{-10}{2}$$
$$x = -5$$

EXAMPLE

If the equation contains parentheses, use the distributive property to remove them.

Solve $4(x - 6) = 40$.

$$4(x - 6) = 40$$
$$4x - 24 = 40$$
$$4x - 24 + 24 = 40 + 24$$
$$4x = 64$$
$$\frac{4x}{4} = \frac{64}{4}$$
$$x = 16$$

EXAMPLE

Solve $\frac{x-3}{2} = \frac{2x+4}{5}$.

$$\frac{10}{1} \cdot \frac{(x-3)}{2} = \frac{10}{1} \cdot \frac{(2x+4)}{5}$$
$$5(x-3) = 2(2x+4)$$
$$5x - 15 = 4x + 8$$
$$5x - 15 - 4x = 4x + 8 - 4x$$
$$x - 15 = 8$$
$$x - 15 + 15 = 8 + 15$$
$$x = 23$$

When you have fractional terms, you can eliminate fractions by multiplying each term on both sides of the equation by the least common denominator. Remember: The fraction bar is a grouping symbol, so enclose numerators in parentheses as needed.

EXERCISE 20.2

For 1 to 18, solve the equation.

1. $\frac{3}{4} + x = 1.25$

2. $2x = 30$

3. $6x - 3 = 13$

4. $-9.22 + x = 10.25 - 8.43$

5. $6x + 25 = 2x + 5$

6. $x + 3(x - 2) = 2x - 4$

7. $\frac{x+3}{5} = \frac{x-1}{2}$

8. $3x + 2 = 6x - 4$

9. $9 + x + 3 = 18 - (x - 4)$

10. $\frac{x}{6} - \frac{1}{2} = \frac{x}{3}$

11. $\frac{2x}{3} + \frac{5x}{6} + \frac{x}{4} = 42$

12. $\frac{5x}{8} - \frac{7x}{8} + \frac{x}{3} = 4$

13. $\frac{x+1}{2} = \frac{x+2}{4} + \frac{x+4}{6}$

14. $\frac{x-1}{12} - \frac{x+1}{16} = \frac{1}{24}$

15. $0.05x = 42$

16. $x + 0.08x = 90.72$

17. $x + 0.7 = 5.4$

18. $0.06x - 0.25 = 0.03x + 0.35$

For 19 to 20, write an equation and solve.

19. If one half of x is added to one-third of x, the sum is 5. Find x.

20. If x is added to both the numerator and denominator of the fraction $\frac{5}{11}$ to yield the fraction $\frac{3}{5}$, what is the value of x?

Solving Linear Equations with Two or More Variables for a Specific Variable

Use the tools for solving a one-variable linear equation to solve a linear equation with two (or more) variables, such as $6x + 2y = 10$, for one of the variables in terms of the other variable(s). As you solve for the variable of interest, simply treat the other variable(s) in the equation as constant(s).

When solving $6x + 2y = 10$ for y, treat $6x$ as if it were a constant.

EXAMPLE

Solve $6x + 2y = 10$ for y.

Solution:

$$\begin{aligned} 6x + 2y &= 10 \\ 6x + 2y - 6x &= 10 - 6x \\ 2y &= 10 - 6x \\ \frac{2y}{2} &= \frac{10 - 6x}{2} \\ y &= 5 - 3x \\ y &= -3x + 5 \end{aligned}$$

EXAMPLE

Solve $-12x + 6y = 9$ for y.

Solution:

$$\begin{aligned} -12x + 6y &= 9 \\ -12x + 6y + 12x &= 9 + 12x \\ 6y &= 9 + 12x \\ \frac{6y}{6} &= \frac{9 + 12x}{6} \\ y &= \frac{3}{2} + 2x \\ y &= 2x + \frac{3}{2} \end{aligned}$$

EXAMPLE

Solve $P = 2(L + W)$ for L.

Solution:

$$\begin{aligned} P &= 2(L+W) \\ P &= 2L + 2W \\ 2L + 2W &= P \\ 2L + 2W - 2W &= P - 2W \\ 2L &= P - 2W \\ \frac{2L}{2} &= \frac{P - 2W}{2} \\ L &= \frac{P - 2W}{2} \end{aligned}$$

For convenience, you can switch the sides of an equation.

EXERCISE 20.3

Solve for the variable indicated.

1. $\frac{3}{4}x = 12y$, for x.
2. $8y = 12x$, for y.
3. $6x - 4 = 8y$, for y.
4. $y - 3 = x$, for y.
5. $7y = 2y - (x - 6y)$, for x.
6. $C = 2\pi r$ for r.
7. $V = Bh$, for h.
8. $F = \frac{9}{5}C + 32$, for C.
9. $P = 2(L + W)$, for W.
10. $\frac{x}{6} - \frac{1}{2} = -\frac{y}{3}$, for y.

Solving One-Variable Linear Inequalities

If you replace the equal sign in a linear equation with $<$, $>$, $\leq$, or $\geq$, the result is a **linear inequality**. Solve linear inequalities just about the same way you solve equations. There is just one important difference: When you multiply or divide both sides of an inequality by a *negative* number, you must *reverse* the direction of the inequality symbol.

EXAMPLE

Solve $-3x - 7 < 14$.

$$\begin{aligned} -3x - 7 &< 14 \\ -3x - 7 + 7 &< 14 + 7 \\ -3x &< 21 \\ \frac{-3x}{-3} &> \frac{21}{-3} \\ x &> -7 \end{aligned}$$

EXAMPLE

Solve $4(x - 6) \geq 44$.

$$\begin{aligned} 4(x - 6) &\geq 44 \\ 4x - 24 &\geq 44 \\ 4x - 24 + 24 &\geq 44 + 24 \\ 4x &\geq 68 \\ \frac{4x}{4} &\geq \frac{68}{4} \\ x &\geq 17 \end{aligned}$$

EXERCISE 20.4

Solve the inequality.

1. $\frac{3}{4} + x > 1.25$

2. $-2x \geq -30$

3. $6x - 3 > 13$

4. $-9.22 + x \leq 10.25 - 8.43$

5. $6x + 25 < 2x + 5$

6. $x + 3(x - 2) > 2x - 4$

7. $\frac{x+3}{5} \leq \frac{x-1}{2}$

8. $3x + 2 > 6x - 4$

9. $9 + x + 3 \geq 18 - (x - 4)$

10. $\frac{x}{6} - \frac{1}{2} > \frac{x}{3}$

11. $\frac{2x}{3} + \frac{5x}{6} + \frac{x}{4} < 42$

12. $\frac{5x}{8} - \frac{7x}{8} + \frac{x}{3} \leq 4$

13. $\frac{x+1}{2} \geq \frac{x+2}{4} + \frac{x+4}{6}$

14. $\frac{x-1}{12} - \frac{x+1}{16} > \frac{1}{24}$

15. $0.05x < 42$

CHAPTER 21

One-Variable Quadratic Equations

Basic Concepts

A **quadratic equation** in the variable x is one that can be written in the standard form $ax^2 + bx + c = 0$, where a, b, and c are constants with $a \neq 0$.

Notice that the nonzero leading term in a quadratic equation has degree 2, and that in standard form, one side of the equation is zero.

EXAMPLE

- $3x^2 - 2x + 11 = 0$ is a quadratic equation in standard form.
- $x^2 - 7 = 0$ is a quadratic equation in standard form.

If a quadratic equation is not in standard form, you can use algebraic manipulation to write it in standard form.

EXAMPLE

- In standard form, $x^2 + x = 6$ is $x^2 + x - 6 = 0$.
- In standard form, $-4x + 4 = -x^2$ is $x^2 - 4x + 4 = 0$.
- In standard form, $(2x - 3)(x + 4) = 0$ is $2x^2 + 5x - 12 = 0$.

Quadratic equations are always solvable for the variable given; however, in some instances, the solution will yield complex numbers, not real numbers. The solution values that make the quadratic equation true are its **roots**. For the purpose of this book, the discussion of quadratic equations will be restricted to real number roots. When you are instructed to solve the equation $ax^2 + bx + c = 0$, find all *real* values for x that make the equation true.

EXERCISE 21.1

For 1 to 5, write the quadratic equation in standard form.

1. $2 + x^2 = 3x$

2. $-6x + 9 = -x^2$

3. $(x - 3)(x - 5) = 0$

4. $10x = 25x^2 + 1$

5. $(3x - 1)(2x - 5) = 0$

For 6 to 15, state whether the equation is true or false for the given value of the variable.

6. $2 + x^2 = 3x,\ x = 2$

7. $-6x + 9 = -x^2,\ x = 3$

8. $(x - 3)(x - 5) = 0,\ x = -5$

9. $12x = 36x^2 + 1,\ x = \frac{1}{6}$

10. $(3x - 1)(2x - 5) = 0,\ x = \frac{5}{2}$

11. $2 + x^2 = 3x,\ x = 1$

12. $-6x + 9 = -x^2, x = -3$

13. $(x - 3)(x - 5) = 0,\ x = 5$

14. $(x - 3)(x - 5) = 0,\ x = 3$

15. $(3x - 1)(2x - 5) = 0,\ x = \frac{1}{3}$

Solving Quadratic Equations of the Form $ax^2 + c = 0$

Normally, the first step in solving a quadratic equation is to put it in standard form. However, if there is no x term (that is, if the coefficient b is zero), then solve for x^2, and thereafter take the square root of both sides.

EXAMPLE

Solve $x^2 - 16 = 0$.

$$x^2 - 16 = 0$$
$$x^2 - 16 + 16 = 0 + 16$$
$$x^2 = 16$$
$$x = 4 \text{ or } x = -4$$

Note: A solution such as $x = 4$ or $x = -4$ is usually written $x = \pm 4$.

EXAMPLE

Solve $x^2 = 7$.

$$x^2 = 7$$
$$x = \pm\sqrt{7}$$

EXERCISE 21.2

Solve.

1. $x^2 = 144$

2. $x^2 - 60 = 4$

3. $x^2 - 32 = 0$

4. $9x^2 = 5x^2 + 28$

5. $81 - 4x^2 = 2x^2 + 9$

6. $x^2 = \frac{1}{4}$

7. $\frac{2x^2 - 3}{2} = \frac{4x^2 - 5}{3}$

8. $x^2 - \frac{9}{16} = 0$

9. $5x^2 = 1$

10. $16x^2 = 48$

11. $3x^2 = 30$

12. $180 = 5x^2$

13. $2x^2 - 68 = 0$

14. $4x^2 = -3x^2 + 175$

15. $112 = 5x^2 - x^2$

16. $(x + 4)(x - 4) = 84$

17. $(x + 3)(x - 2) = x$

18. $x\left(\frac{x}{8}\right) = 2$

19. $\frac{(x + 3)(x - 3)}{2} = 8$

20. $\frac{(x - 4)(x + 7)}{3} = x$

Solving Quadratic Equations by Factoring

When you solve a quadratic equation by factoring, first make sure the equation is in standard form. Then factor the nonzero side. Next, apply the **zero factor property** that states the following: If the product of two numbers is zero, then at least one of the numbers is zero. Accordingly, set each factor equal to zero. Finally, solve the two resulting linear equations for x.

EXAMPLE

Solve $x^2 + 2x = 0$ by factoring.

$$x^2 + 2x = 0$$
$$x(x + 2) = 0$$
$$x = 0 \text{ or } x + 2 = 0$$
$$x = 0 \text{ or } x = -2$$

EXAMPLE

Solve $x^2 + x = 6$ by factoring.

$$x^2 + x = 6$$
$$x^2 + x - 6 = 6 - 6$$
$$x^2 + x - 6 = 0$$
$$(x + 3)(x - 2) = 0$$
$$x + 3 = 0 \text{ or } x - 2 = 0$$
$$x = -3 \text{ or } x = 2$$

EXERCISE 21.3

Solve by factoring.

1. $x^2 + 3x - 18 = 0$
2. $x^2 - 9 = 0$
3. $x^2 - 7x = 0$
4. $4x^2 + 5x = 0$
5. $x^2 + 9 = 6x$
6. $x^2 + 7x = 18$
7. $2x^2 + 7x + 6 = 0$
8. $x(x - 5) = -4$
9. $2x^2 - 8 = 3x^2 + 6x$
10. $x^2 - 4x = 12$

11. $x\left(\frac{x}{3}-1\right)=6$

12. $49x^2 = 81$

13. $x^2 + 16x + 5 = 41$

14. $0 = x^2 + 6x + 8$

15. $x(x + 5) = 4(x + 3)$

16. $x(x - 3) = 10$

17. $(x + 3)(3 - x) = 7 + x$

18. $x^2 + 15 = 8x$

19. $3x^2 + 8 = 2x^2 - 6x$

20. $10x - 24 = -x^2$

Solving Quadratic Equations by Completing the Square

When you use the technique of completing the square to solve quadratic equations, begin by rewriting the equation in the form $x^2 + \frac{b}{a}x = -\frac{c}{a}$. Next, add $\left(\frac{1}{2}\cdot\frac{b}{a}\right)^2$ to both sides of the equation, and then express the variable side as a square. Take the square root of both sides, and solve the two resulting linear equations for x.

EXAMPLE

Solve $x^2 + 2x - 24 = 0$ by completing the square.

$$\begin{aligned} x^2 + 2x - 24 &= 0 \\ x^2 + 2x - 24 + 24 &= 0 + 24 \\ x^2 + 2x &= 24 \\ x^2 + 2x + 1 &= 24 + 1 \\ x^2 + 2x + 1 &= 25 \\ (x + 1)^2 &= 25 \\ x + 1 &= 5 \text{ or } x + 1 = -5 \\ x &= 4 \text{ or } x = -6 \end{aligned}$$

EXAMPLE

Solve $x^2 - 2x = 6$ by completing the square.

$$x^2 - 2x = 6$$
$$x^2 - 2x + 1 = 6 + 1$$
$$x^2 + 2x + 1 = 7$$
$$(x + 1)^2 = 7$$
$$x + 1 = \sqrt{7} \text{ or } x + 1 = -\sqrt{7}$$
$$x = -1 + \sqrt{7} \text{ or } x = -1 - \sqrt{7}$$

EXAMPLE

Solve $2x^2 + 8x + 3 = 0$ by completing the square.

$$2x^2 + 8x + 3 = 0$$
$$2x^2 + 8x + 3 - 3 = 0 - 3$$
$$2x^2 + 8x = -3$$
$$\frac{2x^2}{2} + \frac{8x}{2} = \frac{-3}{2}$$
$$x^2 + 4x = -\frac{3}{2}$$
$$x^2 + 4x + 4 = -\frac{3}{2} + 4$$
$$(x + 2)^2 = \frac{5}{2}$$
$$x + 2 = \sqrt{\frac{5}{2}} \text{ or } x + 2 = -\sqrt{\frac{5}{2}}$$
$$x = -2 + \frac{\sqrt{10}}{2} \text{ or } x = -2 - \frac{\sqrt{10}}{2}$$

EXERCISE 21.4

Solve by completing the square.

1. $x^2 + 3x - 18 = 0$

2. $x^2 - 9 = 0$

3. $x^2 - 7x = 0$

4. $4x^2 + 5x = 0$

5. $x^2 + 9 = 6x$

6. $x^2 + 7x = 18$

7. $2x^2 + 7x + 6 = 0$

8. $x(x - 5) = -4$

9. $2x^2 - 8 = 3x^2 + 6x$

10. $x^2 - 4x = 12$

11. $x^2 + 6x - 40 = 0$

12. $x^2 + 8x + 10 = 0$

13. $(x + 2)^2 = 8x$

14. $2 + 4x = x^2 - 8$

15. $x^2 + 10 = 6x + 6$

Solving Quadratic Equations by Using the Quadratic Formula

The solution of the quadratic equation $ax^2 + bx + c = 0$ is given by the **quadratic formula**: $x = \frac{-b \pm \sqrt{b^2 - 4ac}}{2a}$. The expression under the radical, $b^2 - 4ac$, is called the **discriminant** of the quadratic equation. If $b^2 - 4ac = 0$, there is only one root for the equation. If $b^2 - 4ac > 0$, there are two real number roots. And if $b^2 - 4ac < 0$, there is no real number root. In the latter case, both roots are complex numbers because this solution involves the square root of a negative number.

EXAMPLE

Solve $2x^2 + 2x - 5 = 0$ using the quadratic formula.

$a = 2$, $b = 2$, and $c = -5$

$$x = \frac{-b \pm \sqrt{b^2 - 4ac}}{2a} = \frac{-(2) \pm \sqrt{(2)^2 - 4(2)(-5)}}{2(2)} = \frac{-2 \pm \sqrt{4 + 40}}{4}$$

$$= \frac{-2 \pm \sqrt{44}}{4} = \frac{-2 \pm \sqrt{4(11)}}{4} = \frac{-2 \pm 2\sqrt{11}}{4} = \frac{2(-1 \pm \sqrt{11})}{4} = \frac{-1 \pm \sqrt{11}}{2}$$

The solution is $x = \frac{-1 + \sqrt{11}}{2}$ or $x = \frac{-1 - \sqrt{11}}{2}$.

EXAMPLE

Solve $x^2 - 6x + 9 = 0$ using the quadratic formula.

$a = 1, b = -6$, and $c = 9$

$$x = \frac{-b \pm \sqrt{b^2 - 4ac}}{2a} = \frac{-(-6) \pm \sqrt{(-6)^2 - 4(1)(9)}}{2(1)} = \frac{6 \pm \sqrt{36 - 36}}{2}$$

$$= \frac{6 \pm \sqrt{0}}{2} = \frac{6}{2} = 3$$

The solution is $x = 3$.

EXERCISE 21.5

Solve by using the quadratic formula.

1. $x^2 + 3x - 18 = 0$

2. $x^2 - 9 = 0$

3. $x^2 - 7x = 0$

4. $4x^2 + 5x = 0$

5. $x^2 + 9 = 6x$

6. $x^2 + 7x = 18$

7. $2x^2 + 7x + 6 = 0$

8. $x(x - 5) = -4$

9. $2x^2 - 8 = 3x^2 + 6x$

10. $x^2 - 4x = 12$

11. $x^2 + 6x - 40 = 0$

12. $x^2 + 8x + 10 = 0$

13. $(x + 2)^2 = 8x$

14. $2 + 4x = x^2 - 8$

15. $x^2 + 10 = 6x + 6$

16. $(x + 4)(x - 4) = 84$

17. $(x + 3)(x - 2) = x$

18. $x\left(\frac{x}{8}\right) = 2$

19. $\frac{(x + 3)(x - 3)}{2} = 8$

20. $\frac{(x - 4)(x + 7)}{3} = x$

CHAPTER 22

The Cartesian Coordinate Plane

The Coordinate Plane and Ordered Pairs

The **Cartesian coordinate plane** (also called the ***xy*-plane**) is defined by two real number lines, one horizontal and one vertical, intersecting at right angles at their zero points. The two real number lines are the **coordinate axes**. The **horizontal axis**, commonly the ***x*–axis**, has positive direction to the right, and

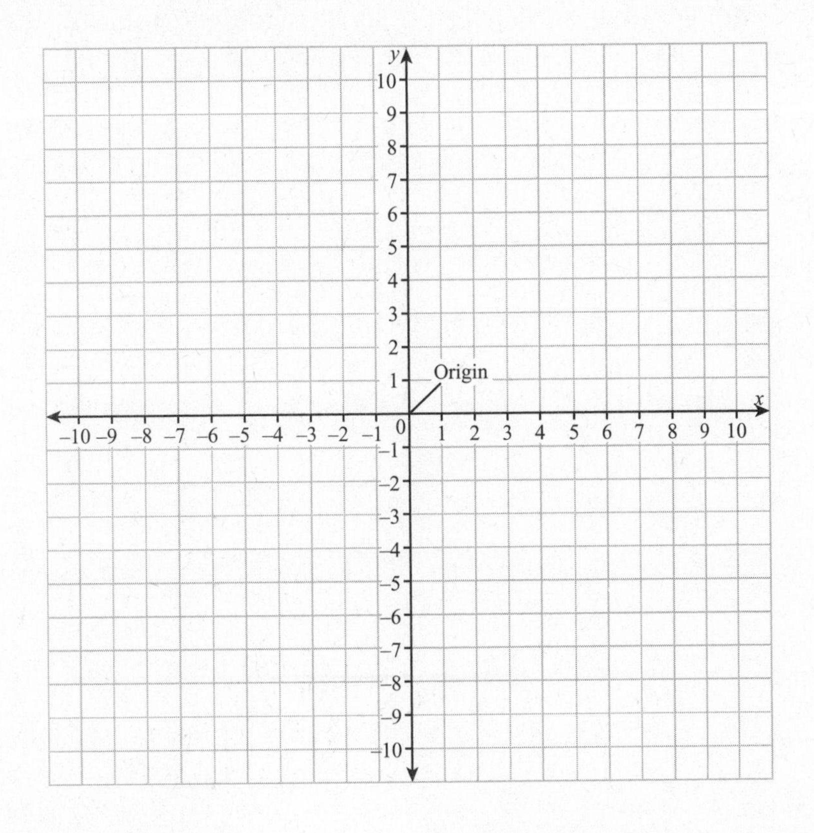

the **vertical axis**, commonly the **y–axis**, has positive direction upward. The two axes determine a plane. Their point of intersection is called the **origin**.

In the coordinate plane, each point P in the plane corresponds to an **ordered pair** (x, y) of real numbers x and y, called its **coordinates**. The ordered pair $(0, 0)$ names the origin. An ordered pair of numbers is written in a definite order so that one number is first and the other is second. The first number is the **x-coordinate** (or **abscissa**), and the second number is the **y-coordinate** (or **ordinate**). The order in the ordered pair (x, y) that corresponds to a point P is important. The first coordinate, x, is the perpendicular horizontal distance (right or left) of the point P from the y-axis. If x is positive, P is to the right of the y-axis, and if x is negative, it is to the left of it. The second coordinate, y, is the perpendicular vertical distance (up or down) of the point P from the x-axis. If y is positive, P is above the x-axis; if y is negative, it is below it.

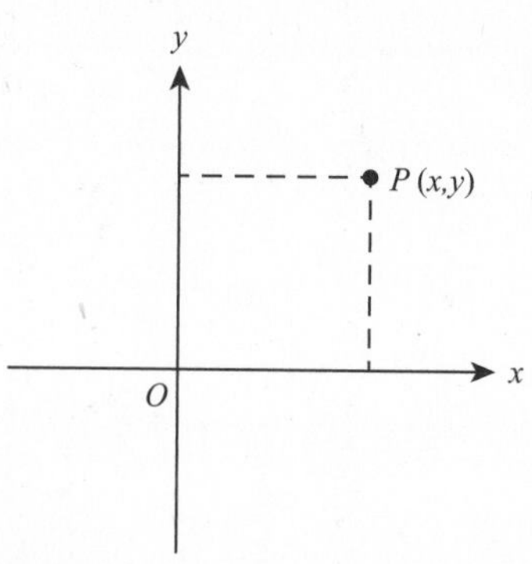

Point P in a Coordinate Plane

EXAMPLE

In the figure shown, the ordered pairs corresponding to points A, B, C, D, and E are $(-7, 4)$, $(-3, 0)$, $(4, 5)$, $(0, -6)$, and $(8, -2)$, respectively.

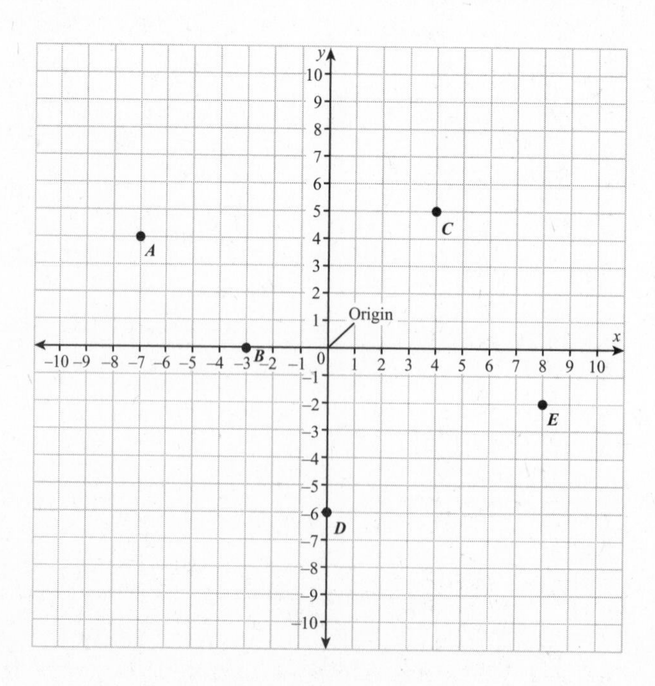

EXERCISE 22.1

In 1 to 5, fill in the blank(s) to make a true statement.

1. The coordinate plane is defined by two ____________ number lines.

2. The ____________ is the intersection of the x–axis and the y–axis.

3. Each point in the coordinate plane corresponds to an ____________ pair of real numbers.

4. In the coordinate plane, the horizontal axis is the ____________ with positive numbers to the ____________ of the origin.

5. In the coordinate plane, the vertical axis is the ____________ with positive numbers ____________ the origin.

In 6 to 15, for each point shown in the coordinate plane, use integer values to write its corresponding ordered pair.

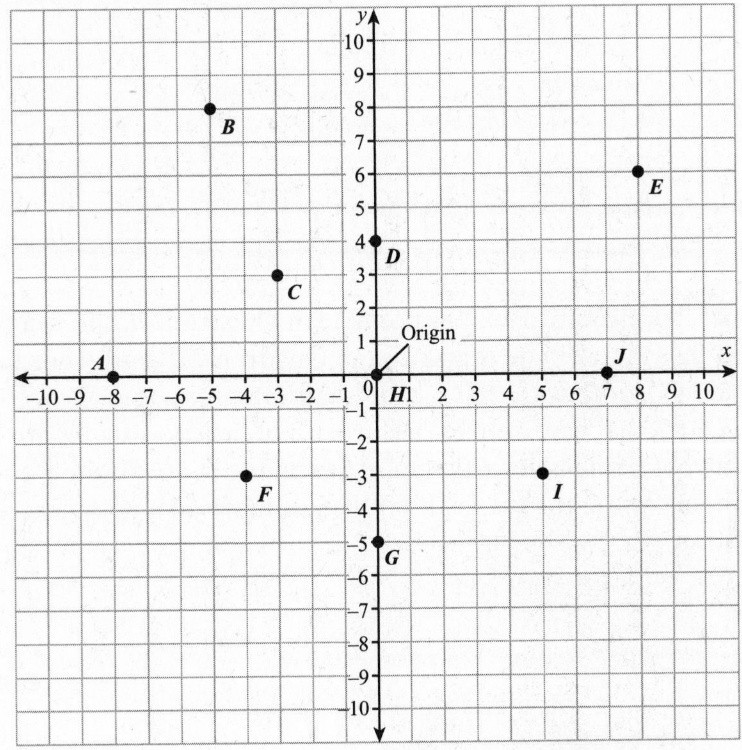

6. A

7. B

8. C

9. D

10. E

11. F

12. G

13. H

14. I

15. J

Quadrants of the Coordinate Plane

The axes divide the coordinate plane into four **quadrants**. The quadrants are numbered with Roman numerals **I, II, III,** and **IV**, beginning in the upper right and going around counterclockwise.

Don't overlook that the quadrants are numbered *counterclockwise*.

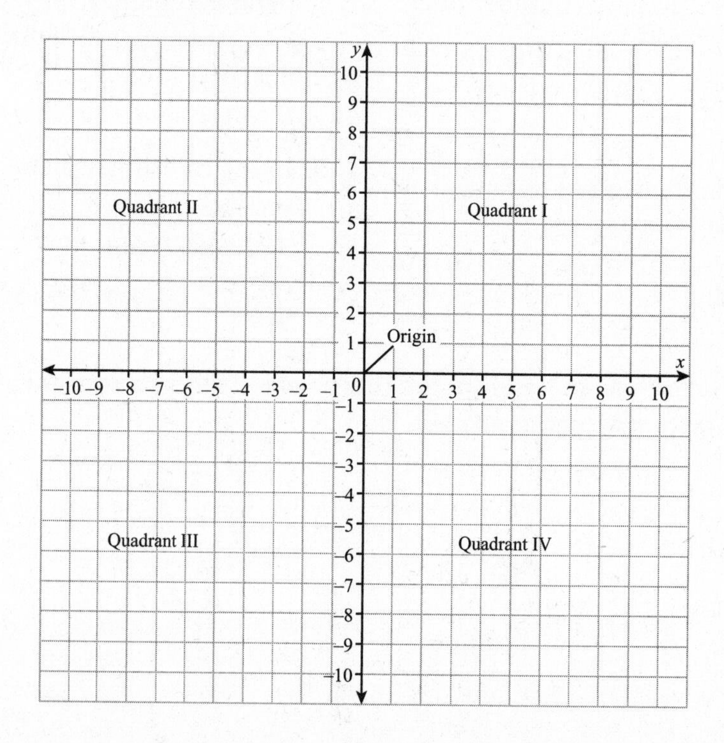

In Quadrant I, both coordinates are positive; in Quadrant II, the x-coordinate is negative and the y-coordinate is positive; in Quadrant III, both coordinates are negative; and in Quadrant IV, the x-coordinate is positive and the y-coordinate is negative. Points that have zero as one or both of the coordinates are on the axes. If the x-coordinate is zero, the point lies on the y-axis. If the y-coordinate is zero, the point lies on the x-axis. If both coordinates of a point are zero, the point is at the origin.

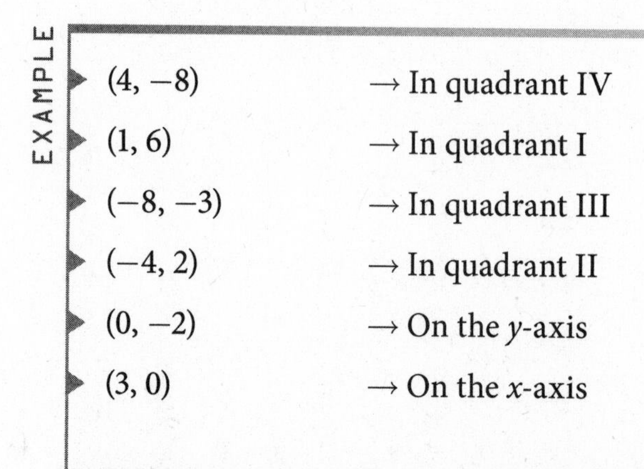
EXAMPLE

- $(4, -8)$ → In quadrant IV
- $(1, 6)$ → In quadrant I
- $(-8, -3)$ → In quadrant III
- $(-4, 2)$ → In quadrant II
- $(0, -2)$ → On the y-axis
- $(3, 0)$ → On the x-axis

EXERCISE 22.2

In 1 to 5, fill in the blank(s).

1. The quadrants in a coordinate plane are numbered in a ____________ direction.
2. In quadrant II, the x-coordinate is ____________.
3. In quadrant IV, the x-coordinate is ____________.
4. If the x-coordinate is zero, the point lies on the ____________.
5. If the y-coordinate is zero, the point lies on the ____________.

In 6 to 20, identify the quadrant in which the point lies or the axis on which it lies.

6. $(1, 3)$
7. $(4, 7)$
8. $(-1, -2)$
9. $(10, 0)$
10. $(-1, 1)$
11. $(6, -5)$
12. $(0, 3)$
13. $(-2.5, 1.5)$
14. $(4.7, 4.8)$
15. $\left(-\frac{1}{2}, -\frac{3}{4}\right)$
16. $\left(\frac{\pi}{2}, 0\right)$
17. $(0, -1)$
18. $\left(\sqrt{7}, -\sqrt{7}\right)$
19. $\left(-\frac{1}{2}, 3\right)$
20. $\left(\frac{2}{3}, 0\right)$

CHAPTER 23

Formulas for the Coordinate Plane

Distance Formula

Use the following formula to find the distance between two points in the coordinate plane.

Distance Between Two Points The distance d between two points (x_1, y_1) and (x_2, y_2) in a coordinate plane is given by $d = \sqrt{(x_2 - x_1)^2 + (y_2 - y_1)^2}$.

To avoid careless errors when using the distance formula, enclose substituted *negative* values in parentheses.

Note that if (x_1, y_1) and (x_2, y_1) are two points that lie on the same horizontal line, an application of the distance formula yields the distance between them as simply $|x_2 - x_1|$. Similarly, if (x_1, y_1) and (x_1, y_2) are two points that lie on the same vertical line, the distance between them is simply $|y_2 - y_1|$.

EXAMPLE

▶ Find the distance between the points (−1, 4) and (5, −3).

Let $(x_1, y_1) = (-1, 4)$ and $(x_2, y_2) = (5, -3)$. Substitute $x_1 = -1$, $y_1 = 4$, $x_2 = 5$, and $y_2 = -3$ into the distance formula:

$$d = \sqrt{(x_2 - x_1)^2 + (y_2 - y_1)^2} = \sqrt{(5 - (-1))^2 + ((-3) - 4)^2}$$
$$= \sqrt{(5 + 1)^2 + (-3 - 4)^2}$$
$$= \sqrt{(6)^2 + (-7)^2} = \sqrt{36 + 49} = \sqrt{85}$$

The distance between (−1, 4) and (5, −3) is $\sqrt{85}$. Note: Leave this answer as a radical.

EXAMPLE

Find the distance between the points (3, 4) and (−8, 4).

Because (3, 4) and (−8, 4) lie on the same horizontal line, the distance between these two points is $|-8-3| = |-11| = 11$.

EXAMPLE

Find the distance between the points (9, 2) and (9, 10).

Because (9, 2) and (9, 10) lie on the same horizontal line, the distance between these two points is $|10-2| = |8| = 8$.

EXERCISE 23.1

Find the distance between the two points. (Write the exact answer in simplest radical form for irrational answers.)

1. (6, −5), (4, 2)

2. (0, 3), (4, −4)

3. (−2, 1), (−2, −7)

4. (7, 8), (5, 1)

5. $\left(-\frac{1}{2}, -\frac{3}{4}\right), \left(-\frac{5}{2}, -\frac{3}{4}\right)$

6. (1, 3), (−1, −3)

7. (4, 7), (−3, 5)

8. (−1, −2), (2, 3)

9. (10, 0), (15, 0)

10. (−1, 1), (−5, −4)

11. (−6, −5), (−2, −3)

12. (5, 3), (−1, −6)

13. (−3, −2), (−1, 4)

14. (8.7, 4.8), (−8.3, 4.8)

15. (0, 0), (6, 8)

16. (−8, 0), (−5, 1)

17. (0, −1), (0, 12)

18. $(\sqrt{7}, -\sqrt{7})$, (0, 0)

19. $\left(-\frac{1}{2}, 3\right), \left(-\frac{3}{2}, 6\right)$

20. (−4, 0), (0, 3)

Midpoint Formula

Use the following formula to find the midpoint between two points in the coordinate plane.

When you use the midpoint formula, be sure to put plus signs, not minus signs, between the two x values and the two y values.

Midpoint Between Two Points The midpoint between two points (x_1, y_1) and (x_2, y_2) in a coordinate plane is the point with coordinates: $\left(\frac{x_1 + x_2}{2}, \frac{y_1 + y_2}{2}\right)$.

EXAMPLE

▶ Find the midpoint between (–1, 4) and (5, –3).

Let $(x_1, y_1) = (-1, 4)$ and $(x_2, y_2) = (5, -3)$. Substitute $x_1 = -1, y_1 = 4$, $x_2 = 5$, and $y_2 = -3$ into the midpoint formula:

$$\text{Midpoint} = \left(\frac{x_1 + x_2}{2}, \frac{y_1 + y_2}{2}\right) = \left(\frac{-1+5}{2}, \frac{4-3}{2}\right) = \left(\frac{4}{2}, \frac{1}{2}\right) = \left(2, \frac{1}{2}\right)$$

The midpoint between (–1, 4) and (5, –3) is $\left(2, \frac{1}{2}\right)$.

EXERCISE 23.2

Find the midpoint between the two points.

1. (6, –5), (4, 2)

2. (0, 3), (4, –4)

3. (–2, 1), (–2, –7)

4. (7, 8), (5, 1)

5. $\left(-\frac{1}{2}, -\frac{3}{4}\right), \left(-\frac{5}{2}, -\frac{3}{4}\right)$

6. (1, 3), (–1, –3)

7. (4, 7), (–3, 5)

8. (–1, –2), (2, 3)

9. (10, 0), (15, 0)

10. (–1, 1), (–5, –4)

11. (–6, –5), (–2, –3)

12. (5, 3), (–1, –6)

13. (–3, –2), (–1, 4)

14. (8.7, 4.8), (–8.3, 4.8)

15. (0, 0), (6, 8)

16. (–8, 0), (–5, 1)

17. (0, –1), (0, 12)

18. $(\sqrt{7}, -\sqrt{7})$, (0, 0)

19. $\left(-\frac{1}{2}, 3\right), \left(-\frac{3}{2}, 6\right)$

20. (–4, 0), (0, 3)

Slope Formula

Two points are distinct if they are not the same point.

When you use the slope formula, be sure to subtract the coordinates in the same order in both the numerator and denominator. That is, x_2 is the first term in the numerator, so y_2 must be the first term in the denominator. It is also a good idea to enclose substituted *negative* values in parentheses—to guard against careless errors.

You can construct a line through two distinct points in a coordinate plane. The **slope** of a *nonvertical* line through two points describes the steepness or slant (if any) of the line. To calculate the slope of a nonvertical line, use the following formula.

Slope of a Line Through Two Points The slope m of a nonvertical line through two distinct points, (x_1, y_1) and (x_2, y_2), is given by $m = \frac{y_2 - y_1}{x_2 - x_1}$, provided $x_1 \neq x_2$.

From the formula, you can see that the slope is the ratio of the change in vertical coordinates (the *rise*) to the change in horizontal coordinates (the *run*). Thus, Slope $= \frac{\text{Rise}}{\text{Run}}$. The following figure illustrates the rise and run for the slope of the line through points $P_1(x_1, y_1)$ and $P_2(x_2, y_2)$.

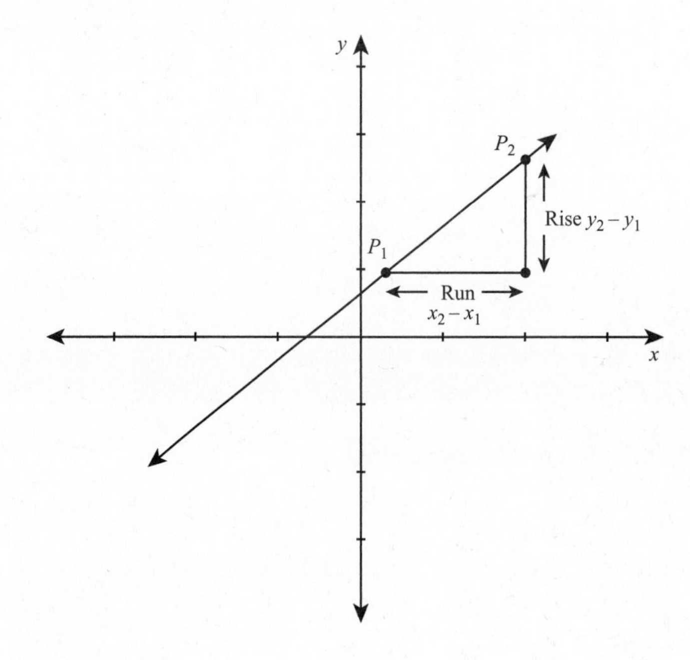

Rise and Run

Lines that slant upward from left to right have positive slopes, and lines that slant downward from left to right have negative slopes. Horizontal lines have zero slope; however, slope for vertical lines is undefined.

EXAMPLE

Find the slope of the line through the points shown.

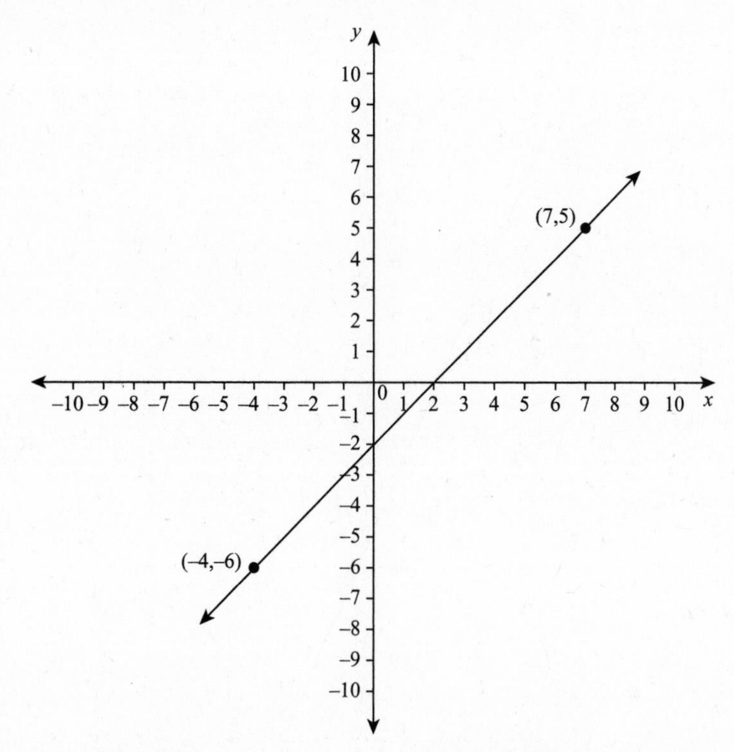

Let $(x_1, y_1) = (7, 5)$ and $(x_2, y_2) = (-4, -6)$. Substitute $x_1 = 7$, $y_1 = 5$, $x_2 = -4$, and $y_2 = -6$ into the slope formula:

$$m = \frac{y_2 - y_1}{x_2 - x_1} = \frac{(-6) - 5}{(-4) - 7} = \frac{-6 - 5}{-4 - 7} = \frac{-11}{-11} = 1$$

The slope of the line through the points (7, 5) and (−4, −6) is 1. Note that the line slants upward from left to right—so the slope should be positive.

EXAMPLE

Find the slope of the line through the points shown.

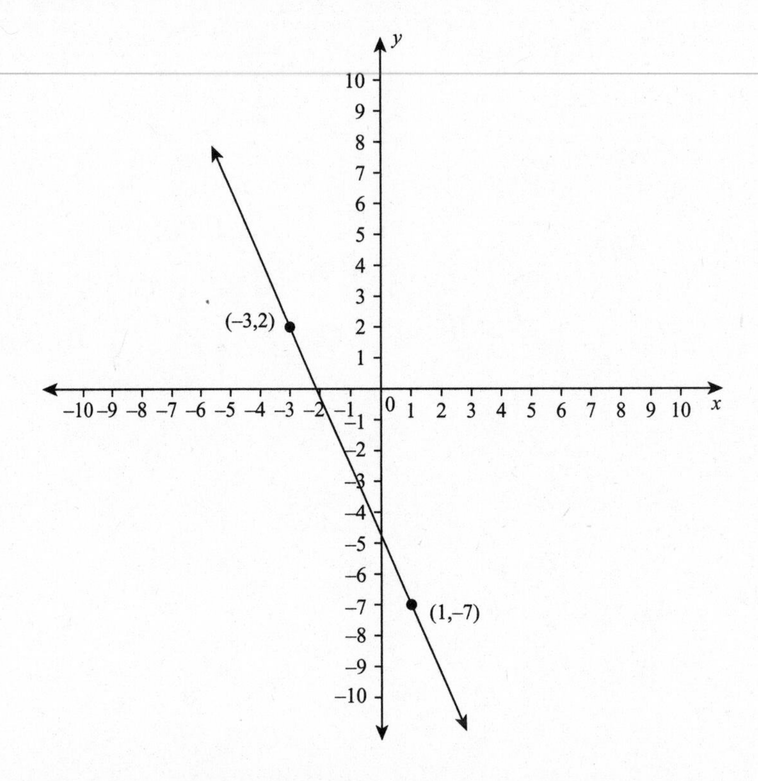

Let $(x_1, y_1) = (-3, 2)$ and $(x_2, y_2) = (1, -7)$. Substitute $x_1 = -3$, $y_1 = 2$, $x_2 = 1$, and $y_2 = -7$ into the slope formula:

$$m = \frac{y_2 - y_1}{x_2 - x_1} = \frac{(-7) - 2}{1 - (-3)} = \frac{-7 - 2}{1 + 3} = \frac{-9}{4} = -\frac{9}{4}$$

The slope of the line through the points $(-3, 2)$ and $(1, -7)$ is $-\frac{9}{4}$.

Note that the line slants downward from left to right—so the slope should be negative.

EXAMPLE

Find the slope of the line through the points (3, 4) and (–8, 4).

Because (3, 4) and (–8, 4) lie on the same horizontal line, the slope of the line is zero.

EXAMPLE

Find the slope of the line through the points (9, 2) and (9, 10).

Because (9, 2) and (9, 10) lie on the same vertical line, the slope of the line is undefined.

If two lines are parallel, their slopes are equal; and if two lines are perpendicular, their slopes are negative reciprocals of each other.

EXAMPLE

Find the slope m_1 of a line that is parallel to the line through (–3, 4) and (–1, –2).

Given that two parallel lines have equal slopes, then m_1 will equal the slope m of the line through (–3, 4) and (–1, –2); that is, $m_1 = m$.

$$m = \frac{y_2 - y_1}{x_2 - x_1} = \frac{(-2) - 4}{(-1) - (-3)} = \frac{-2 - 4}{-1 + 3} = \frac{-6}{2} = -3$$

Therefore, $m_1 = m = -3$.

EXAMPLE

Find the slope m_2 of a line that is perpendicular to the line through (–4, 2) and (1, –8).

Given that the slopes of two perpendicular lines are negative reciprocals of each other, then m_2 equals the negative reciprocal of the slope m of the line through (–4, 2) and (1, –8); that is, $m_2 = -\frac{1}{m}$.

$$m = \frac{y_2 - y_1}{x_2 - x_1} = \frac{(-8) - 2}{1 - (-4)} = \frac{-8 - 2}{1 + 4} = \frac{-10}{5} = -2$$

Therefore, $m_2 = -\frac{1}{m} = -\frac{1}{-2} = \frac{1}{2}$.

EXERCISE 23.3

In 1 to 5, fill in the blank to make a true statement.

1. Lines that slant downward from left to right have ____________ slopes.

2. Lines that slant upward from left to right have ____________ slopes.

3. Horizontal lines have ____________ slope.

4. The slope of a vertical line is ____________.

5. The slope of a line that is perpendicular to a line that has slope $\frac{2}{3}$ is ____________.

In 6 to 16, find the slope of the line between the two points.

6. $(6, -5), (4, 2)$

7. $(0, 3), (4, -4)$

8. $(-2, 1), (-5, -7)$

9. $(7, 8), (5, 1)$

10. $\left(-\frac{1}{2}, -\frac{3}{4}\right), \left(-\frac{5}{2}, -\frac{3}{4}\right)$

11. $(1, 3), (-1, -3)$

12. $(4, 7), (-3, 5)$

13. $(-1, -2), (2, 3)$

14. $(10, 0), (15, 0)$

15. $(-1, 1), (-5, -4)$

16. $(-6, -5), (-2, -3)$

In 17 to 20, find the indicated slope.

17. Find the slope of a line that is parallel to the line through $(-1, 4), (-2, -3)$.

18. Find the slope of a line that is perpendicular to the line through $(-1, 4), (-2, -3)$.

19. Find the slope of a line that is parallel to the line through $(0, -8), (-3, 5)$.

20. Find the slope of a line that is perpendicular to the line through $(0, -8), (-3, 5)$.

CHAPTER 24

Graphing Lines in the Plane

Graphing $y = mx + b$

The equation $y = mx + b$ is the **slope-intercept form** of the equation of a line. The graph of $y = mx + b$ is a straight line that has **slope** m and **y intercept** b.

The number b is the y-value of the point on the line where it crosses the y-axis. Thus, the intersection point has coordinates $(0, b)$.

EXAMPLE

Graph $y = 3x + 1$.

The line has slope 3 and y intercept 1. First, plot $(0, 1)$, the point at which the line crosses the y-axis. Next, using that the slope $\frac{3}{1} = \frac{\text{Rise}}{\text{Run}}$, move 1 unit right and 3 units up to plot a second point on the line. Then construct a line through the two plotted points.

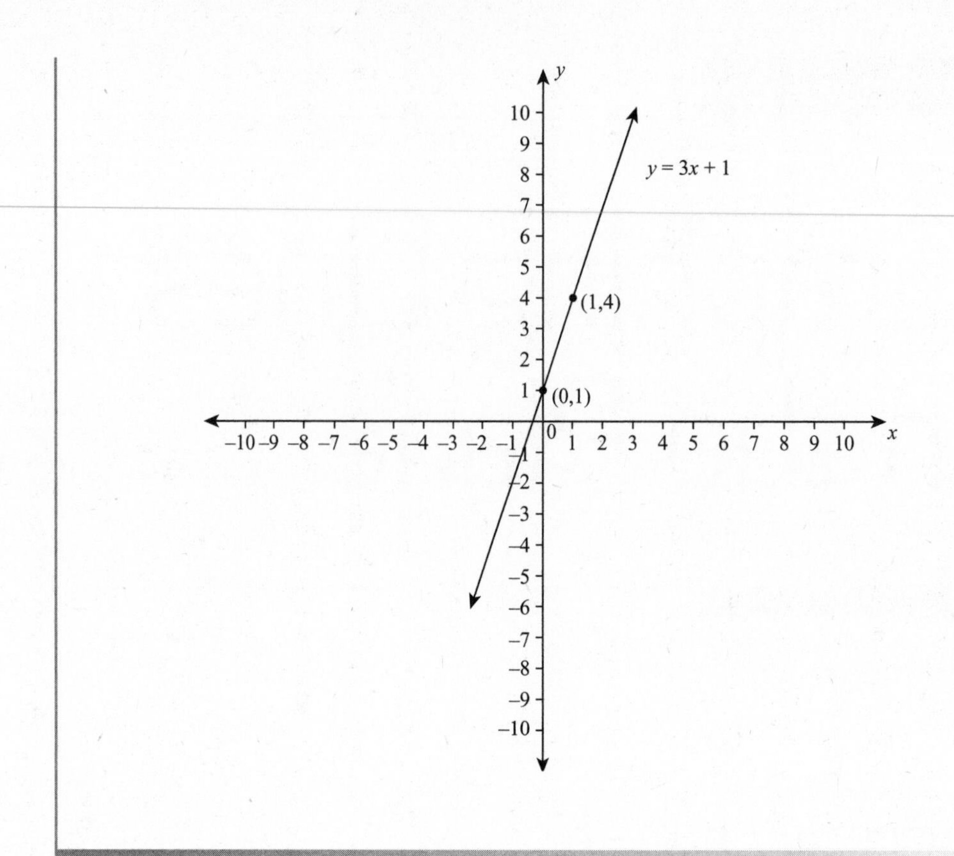

EXAMPLE

▶ Graph $y = -\frac{2}{3}x + 5$.

The line has slope $-\frac{2}{3}$ and y intercept 5. First, plot (0, 5), the point at which the line crosses the y-axis. Next, using that the slope $\frac{-2}{3} = \frac{\text{Rise}}{\text{Run}}$, move 3 units right and 2 units down to plot a second point on the line. Then construct a line through the two plotted points.

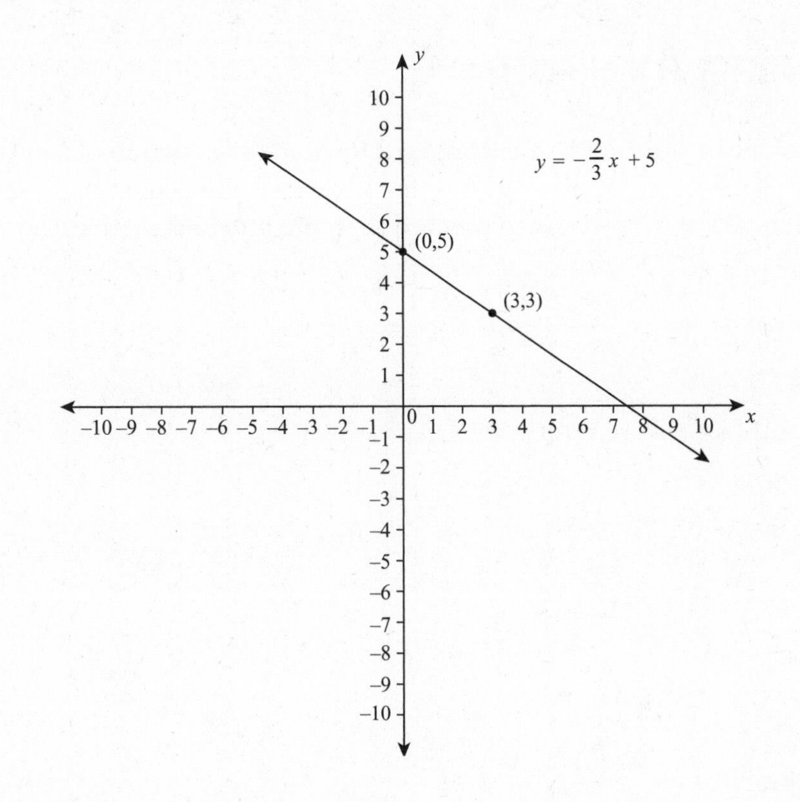

EXERCISE 24.1

For 1 to 5, state (a) the slope and (b) the y-intercept of the graph of the equation.

1. $y = -3x + 4$

2. $y = \frac{3}{4}x - 1$

3. $y = 6x$

4. $y = -\frac{4}{3}x - \frac{1}{2}$

5. $y = -x + 3$

For 6 to 15, graph the equation.

6. $y = -3x + 4$

7. $y = \frac{3}{4}x - 1$

8. $y = 6x$

9. $y = -\frac{4}{3}x - \frac{1}{2}$

10. $y = -\frac{4}{3}x$

11. $y = -\frac{1}{2}x - 4$

12. $y = -4x + 6$

13. $y = 2x - 4$

14. $y = 5x + 1$

15. $y = \frac{3}{5}x - 6$

Graphing $Ax + By = C$

The equation $Ax + By = C$ is the **standard form** of the equation of a line. For graphing purposes, solve $Ax + By = C$ for y to express the equation in slope-intercept form $y = mx + b$. Then proceed as in the previous section.

EXAMPLE

Graph $2x - 3y = 6$.

First, solve $2x - 3y = 6$ for y:

$$\begin{aligned} 2x - 3y &= 6 \\ 2x - 3y - 2x &= 6 - 2x \\ -3y &= -2x + 6 \\ \frac{-3y}{-3} &= \frac{-2x + 6}{-3} \\ y &= \frac{2}{3}x - 2 \end{aligned}$$

Next, graph $y = \frac{2}{3}x - 2$.

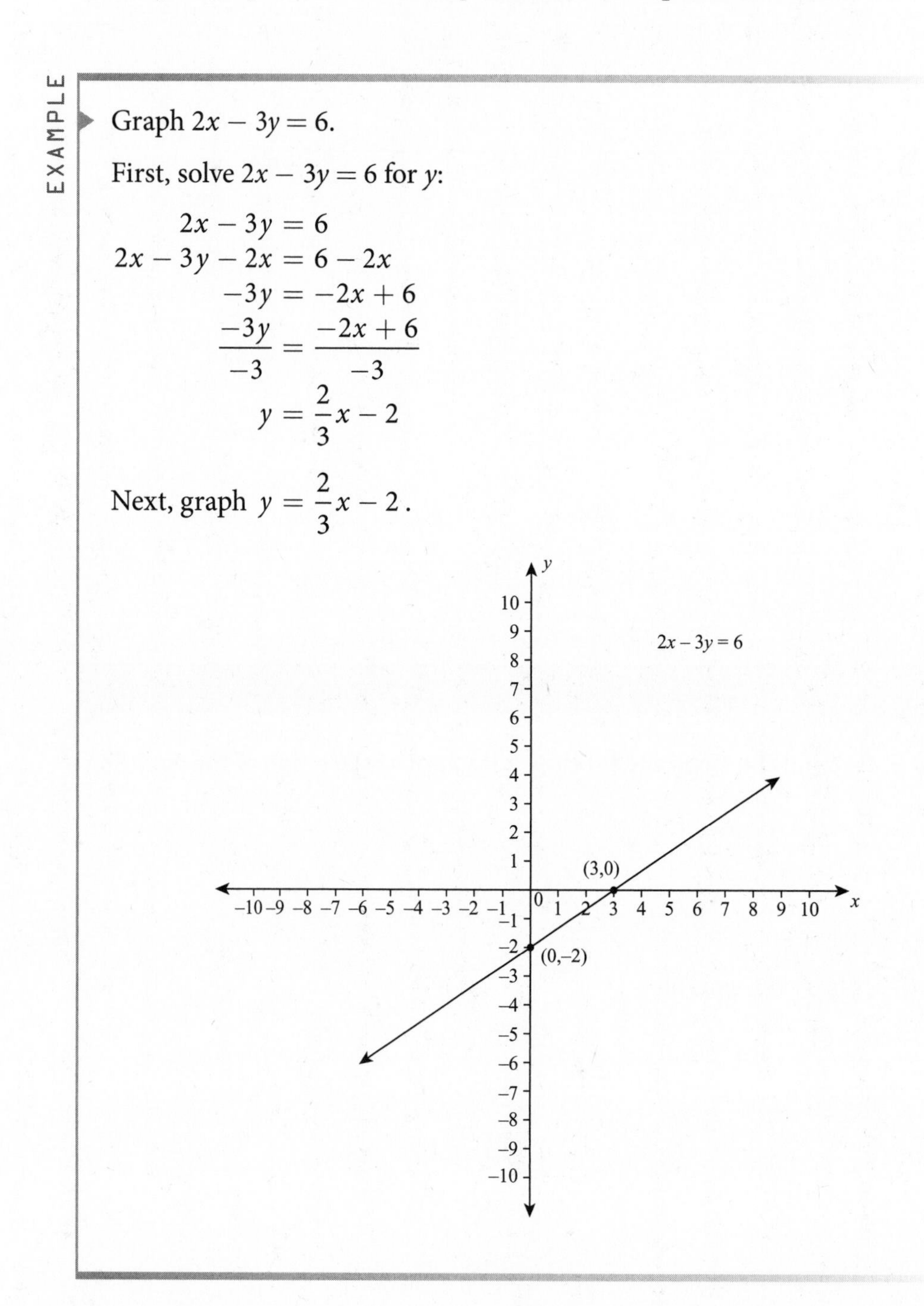

EXAMPLE

Graph $3x + 5y = 15$.

First, solve $3x + 5y = 15$ for y:

$$\begin{aligned} 3x + 5y &= 15 \\ 3x + 5y - 3x &= 15 - 3x \\ 5y &= -3x + 15 \\ \frac{5y}{5} &= \frac{-3x + 15}{5} \\ y &= -\frac{3}{5}x + 3 \end{aligned}$$

Next, graph $y = -\frac{3}{5}x + 3$.

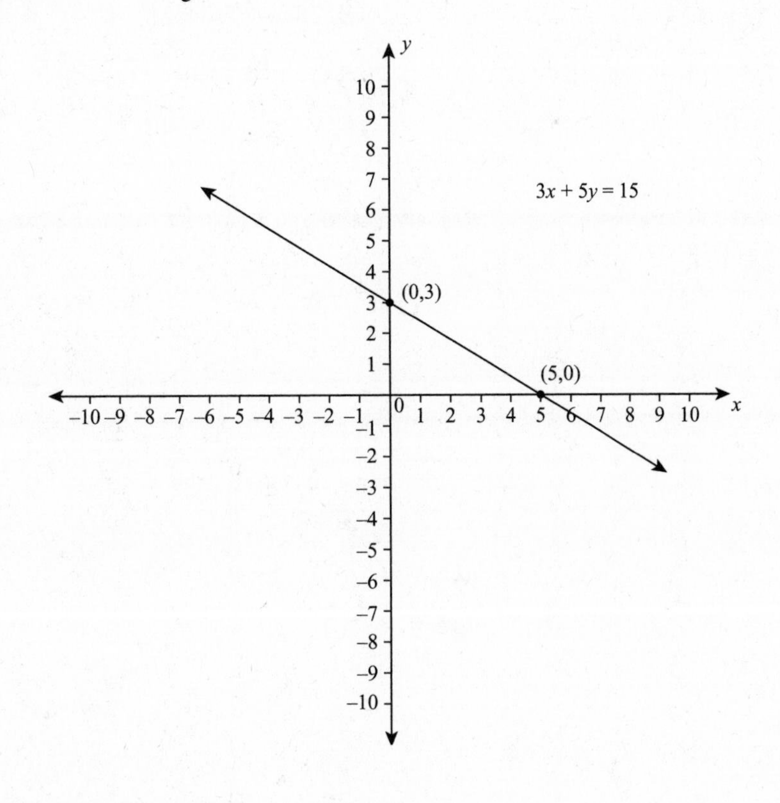

EXERCISE 24.2

For 1 to 5, state (a) the slope and (b) the y-intercept of the graph of the equation.

1. $x - 3y = 9$
2. $3x + y = 6$
3. $x + y = 5$
4. $5x + y = -4$
5. $x - y = 0$

For 6 to 15, graph the equation.

6. $x - 3y = 9$
7. $3x + y = 6$
8. $x + y = 5$
9. $5x + y = -4$
10. $x - y = 0$
11. $3x - 2y = -6$
12. $5x + 4y = 2$
13. $28x - 7y = 21$
14. $x + 5y = 10$
15. $x - 2y + 3 = 0$

CHAPTER 25

Determining the Equation of a Line

Determining the Equation of a Line Given the Slope and y-Intercept

When you are given the slope m of a line and the y-intercept b, simply substitute the given information into $y = mx + b$ to determine the equation of the line.

EXAMPLE

Write the equation of the line that has slope $m = 3$ and y-intercept $= 5$.

The equation of the line is $y = 3x + 5$.

EXAMPLE

Write the equation of the line that has slope $m = \frac{1}{2}$ and y-intercept $= -2$.

The equation of the line is $y = \frac{1}{2}x - 2$.

EXERCISE 25.1

Write the equation of the line using the given information.

1. $m = -1$, y-intercept $= -5$

2. $m = \frac{1}{2}$, y-intercept $= 4$

3. $m = 2$, y-intercept $= -5$

4. $m = -\frac{2}{3}$, y-intercept $= \frac{5}{3}$

5. $m = 4$, y-intercept $= 0$

6. $m = -\frac{3}{8}$, y-intercept $= -\frac{7}{8}$

7. $m = 5$, y-intercept $= 2$

8. $m = -2$, y-intercept $= 3$

9. $m = \frac{1}{2}$, y-intercept $= -1$

10. $m = -\frac{3}{4}$, y-intercept $= 0$

11. $m = -4$, y-intercept $= \frac{1}{2}$

12. $m = -\frac{5}{3}$, y-intercept $= -3$

13. $m = \frac{7}{4}$, y-intercept $= -\frac{1}{2}$

14. $m = 0$, y-intercept $= 4$

15. $m = \frac{3}{8}$, y-intercept $= -7$

Determining the Equation of a Line Given the Slope and One Point on the Line

When you are given the slope m of a line and one point (x_1, y_1) on the line, substitute the given information into the point-slope formula: $m = \frac{y - y_1}{x - x_1}$. Next, solve for y to obtain $y = mx + b$, the slope-intercept form of the equation of the line.

EXAMPLE

Problem

Given the slope $m = 2$ and a point on the line $(3, 5)$, write the equation of the line.

Substitute the given information into the point-slope formula, and then solve for y.

$$m = \frac{y - y_1}{x - x_1}$$

$$2 = \frac{y - 5}{x - 3}$$

$$(y - 5) = 2(x - 3)$$

$$y - 5 = 2x - 6$$

$$y - 5 + 5 = 2x - 6 + 5$$

$$y = 2x - 1$$

The equation of the line is $y = 2x - 1$.

EXAMPLE

Problem

Given the slope $m = \frac{1}{2}$ and a point on the line (–1, 3), write the equation of the line.

Substitute the given information into the point-slope formula, and then solve for y.

$$m = \frac{y - y_1}{x - x_1}$$
$$\frac{1}{2} = \frac{y - 3}{x - (-1)}$$
$$\frac{1}{2} = \frac{y - 3}{x + 1}$$
$$2(y - 3) = 1(x + 1)$$
$$2y - 6 = x + 1$$
$$2y - 6 + 6 = x + 1 + 6$$
$$2y = x + 7$$
$$\frac{2y}{2} = \frac{x + 7}{2}$$
$$y = \frac{1}{2}x + \frac{7}{2}$$

The equation of the line is $y = \frac{1}{2}x + \frac{7}{2}$.

EXERCISE 25.2

Write the equation of the line with the given slope passing through the given point.

1. Slope 2, point (4, 2)
2. Slope 4, point (0, –2)
3. Slope $\frac{1}{2}$, point (–2, –7)
4. Slope –1, point (–3, –4)
5. Slope $\frac{3}{4}$, point (0, 0)
6. Slope –5, point (2, –1)
7. Slope $-\frac{3}{4}$, point (4, 5)
8. Slope –3, point (–7, –3)
9. Slope $-\frac{5}{2}$, point (1, –2)
10. Slope $\frac{3}{8}$, point (–5, 4)
11. Slope $\frac{3}{8}$, point (–2, –3)
12. Slope –3, point (–1, –6)
13. Slope $\frac{1}{2}$, point (–1, 4)
14. Slope $-\frac{4}{3}$, point (3, –5)
15. Slope $\frac{1}{2}$, point (0, 0)

16. Slope 5, point (−5, 1)

17. Slope $-\frac{5}{3}$, (0, −3)

18. Slope −4, $\left(0, \frac{1}{2}\right)$

19. Slope $\frac{7}{4}$, $\left(0, -\frac{1}{2}\right)$

20. Slope 0, (1, 3)

Determining the Equation of a Line Given Two Distinct Points on the Line

When two points are known, you can choose either point to complete the process of writing the equation.

When you are given two distinct points (x_1, y_1) and (x_2, y_2) on a line, first, determine the line's slope by substituting the given information into the slope formula: $m = \frac{y_2 - y_1}{x_2 - x_1}$. Next, proceed as in Lesson 25.2 to use the slope and one of the given points to obtain $y = mx + b$, the slope-intercept form of the equation of the line.

EXAMPLE

Write the equation of the line that contains the points (1, 2) and (3, 4).

First, determine the line's slope.

$$m = \frac{y_2 - y_1}{x_2 - x_1} = \frac{4-2}{3-1} = \frac{2}{2} = 1$$

Next, using $(x_1, y_1) = (1, 2)$, substitute into the point-slope formula, and then solve for y.

$$m = \frac{y - y_1}{x - x_1}$$

$$1 = \frac{y-2}{x-1}$$

$$(y-2) = 1(x-1)$$

$$y - 2 = x - 1$$

$$y - 2 + 2 = x - 1 + 2$$

$$y = x + 1$$

The equation of the line is $y = x + 1$.

EXAMPLE

Write the equation of the line that contains the points (3, −7) and (−1, 4).

First, determine the line's slope.

$$m = \frac{y_2 - y_1}{x_2 - x_1} = \frac{4-(-7)}{(-1)-3} = \frac{4+7}{-4} = -\frac{11}{4}$$

Next, using $(x_1, y_1) = (3, -7)$, substitute into the point-slope formula, and then solve for y.

$$\begin{aligned} m &= \frac{y - y_1}{x - x_1} \\ -\frac{11}{4} &= \frac{y-(-7)}{x-3} \\ -\frac{11}{4} &= \frac{y+7}{x-3} \\ 4(y+7) &= -11(x-3) \\ 4y + 28 - 28 &= -11x + 33 - 28 \\ 4y &= -11x + 5 \\ \frac{4y}{4} &= \frac{-11x+5}{4} \\ y &= -\frac{11}{4}x + \frac{5}{4} \end{aligned}$$

The equation of the line is $y = -\frac{11}{4}x + \frac{5}{4}$.

EXERCISE 25.3

Write the equation of the line that contains the two points.

1. (6, −5), (4, 2)

2. (0, 3), (4, −4)

3. (−2, 1), (−4, −7)

4. (7, 8), (5, 1)

5. $\left(-\frac{1}{2}, -\frac{3}{4}\right), \left(-\frac{5}{2}, -\frac{3}{4}\right)$

6. (1, 3), (−1, −3)

7. (4, 7), (−3, 5)

8. (−1, −2), (2, 3)

9. (10, 0), (15, 0)

10. (−1, 1), (−5, −4)

11. (−6, −5), (−2, −3)

12. (5, 3), (−1, −6)

13. (−3, −2), (−1, 4)

14. (8.7, 4.8), (−8.3, 4.8)

15. (0, 0), (6, 8)

16. (−8, 0), (−5, 1)

17. (1, −1), (0, 12)

18. $(\sqrt{7}, -\sqrt{7})$, (0, 0)

19. $\left(-\frac{1}{2}, 3\right), \left(-\frac{3}{2}, 6\right)$

20. (−4, 0), (0, 3)

CHAPTER 26

Signal Words and Phrases

Common Signal Words and Phrases for Addition

Common signal words and phrases that indicate addition are shown in the following table. Just as in arithmetic, the plus sign (+) is the symbol for addition.

Signal Word or Phrase	Example	Algebraic Symbolism
Plus	x plus 100	$x + 100$
Added to	10 added to an integer n	$n + 10$
Sum	The sum of $2x$ and $3y$	$2x + 3y$
Total of	The total of x, y, and z	$x + y + z$
More than	25 more than $10A$	$10A + 25$
Greater than	5 greater than $8y$	$8y + 5$
Increased by	$4m$ increased by 15	$4m + 15$
Exceeded by	x exceeded by 10	$x + 10$

EXAMPLE

Represent the phrase by an algebraic expression.

The sum of $5x$ and 40 → $5x + 40$
9 more than $6y$ → $6y + 9$
The total of $4x$, $3y$, and $7b$ → $4x + 3y + 7b$
400 increased by $50x$ → $400 + 50x$

EXERCISE 26.1

Represent the phrase by an algebraic expression.

1. The sum of $55x$ and 200
2. 10 more than $5y$
3. The total of $2x$, $8y$, and $9b$
4. 350 increased by $15x$
5. 125 increased by $40\%B$
6. $4x$ plus 5
7. The sum of z and $3z$
8. m exceeded by 15
9. The sum of $3x$ and $4y$
10. 20 added to $5x$
11. 11 more than $9x$
12. 12 exceeded by $2x$
13. The sum of a^2 and b^2
14. 60 greater than $5x$
15. c increased by $10\%c$

Common Signal Words and Phrases for Subtraction

Notice in the table that the order of the terms in subtraction is important.

Common signal words and phrases that indicate subtraction are shown in the following table. Just as in arithmetic, the minus sign (−) is the symbol for subtraction.

Signal Word or Phrase	Examples	Algebraic Symbolism
Minus	x minus 13 13 minus x	$x - 13$ $13 - x$
Subtracted from	40 subtracted from $5K$ $5K$ subtracted from 40	$5K - 40$ $40 - 5K$
Difference	The difference of $2x$ and y The difference of y and $2x$	$2x - y$ $y - 2x$

Less than	z less than 27 27 less than z	$27 - z$ $z - 27$
Fewer than	25 fewer than $2n$ $2n$ fewer than 25	$2n - 25$ $25 - 2n$
Decreased by	y decreased by 50 50 decreased by y	$y - 50$ $50 - y$
Reduced by	370 reduced by $2a$ $2a$ reduced by 370	$370 - 2a$ $2a - 370$
Diminished by	1,000 diminished by B B diminished by 1,000	$1{,}000 - B$ $B - 1{,}000$

EXAMPLE

Represent the phrase by an algebraic expression.

The difference of $50x$ and $75y$ → $50x - 75y$

98 decreased by $4w$ → $98 - 4w$

$17b$ subtracted from 200 → $200 - 17b$

600 reduced by $0.05x$ → $600 - 0.05x$

EXERCISE 26.2

Represent the phrase by an algebraic expression.

1. The difference of $10x$ and $5y$
2. 80 decreased by $2w$
3. $20b$ subtracted from 500
4. 300 reduced by $0.25x$
5. L subtracted from P
6. The difference between c^2 and a^2
7. 200 subtracted from K
8. x minus 13
9. 30 minus y
10. 10 less than $2x$
11. $5x$ diminished by $2x$
12. $7n$ fewer than 12
13. K less than 100
14. 420 reduced by $5y$
15. 8 fewer than $6x$

Common Signal Words and Phrases for Multiplication

The times sign ($\times$) is not used in algebraic expressions.

Common signal words and phrases that indicate multiplication are shown in the following table. Three ways to denote multiplication in algebraic expressions are juxtaposition (that is, side-by-side placement), parentheses, and the dot multiplication symbol ($\cdot$). Commonly, use juxtaposition and parentheses when the factors involve variables. Use the dot multiplication symbol or parentheses when the factors are constants.

Use the word *quantity* to make your meaning clear, as in the quantity $(z + 4)$.

Signal Word or Phrase	Example(s)	Algebraic Symbolism
Times	x times y 8 times 45	xy $8 \cdot 45$ or $(8)(45)$ or $8(45)$
Product	The product of $5m$ and $3n$	$(5m)(3n)$ or $5m(3n)$
Multiplied by	60 multiplied by x	$60x$
Twice, double, triple, etc.	Twice the quantity $(z + 4)$	$2(z + 4)$
Of (when it comes between two numerical quantities)	10 of x 5% of 200 $\frac{3}{4}$ of y	$10x$ $5\%(200)$ $\frac{3}{4}y$
Square, cube, power	The square of x The third power of 5	$x \cdot x = x^2$ $5 \cdot 5 \cdot 5 = 125$

Never use juxtaposition to show multiplication between constant factors. Writing the factors 56 and 8 side by side looks like 568, instead of the product, 56 times 8.

EXAMPLE

Represent the phrase by an algebraic expression.

The product of $50x$ and 20 $\rightarrow (50x)(20)$

56 times 8 $\rightarrow 56 \cdot 8$ or $(56)(8)$

$14x$ multiplied by $3y$ $\rightarrow (14x)(3y)$

25% of B $\rightarrow 25\%B$

EXERCISE 26.3

Represent the phrase by an algebraic expression.

1. The product of $5x$ and y

2. 25 times 3

3. $7x$ multiplied by 8

4. 5% of B

5. y multiplied by 3

6. Twice the quantity $(l + w)$

7. 100 of b

8. $\frac{2}{3}$ of $300x$

9. 0.03 of x

10. d multiplied by π

11. $\frac{1}{2}h$ times the quantity $(b_1 + b_2)$

12. The cube of r

13. The product of $50x$ and 20

14. Twice x^2

15. The fourth power of y

Common Signal Words and Phrases for Division

Signal words and phrases that indicate division are shown in the following table. In algebraic expressions, the fraction bar is used to indicate division as shown here: $\frac{\text{dividend}}{\text{divisor}}$.

Signal Word or Phrase	Examples	Algebraic Symbolism
Divided by	a divided by b b divided by a	$\frac{a}{b}$ $\frac{b}{a}$
Quotient	The quotient of 60 and $5x$ The quotient of $5x$ and 60	$\frac{60}{5x}$ $\frac{5x}{60}$
Ratio	The ratio of W to M The ratio of M to W	$\frac{W}{M}$ $\frac{M}{W}$
For every, for each	x for every 100	$\frac{x}{100}$
Per	$100c$ per m	$\frac{100c}{m}$
Over	x over 5	$\frac{x}{5}$

Observe in the table that the order of the parts in a division expression is important.

Neither of the division symbols, ÷ or $\overline{)\ \ }$, is used in algebraic expressions.

Caution: Division by zero is undefined. For example, $\frac{4x}{0}$ has no meaning.

EXAMPLE

Represent the phrase by an algebraic expression.

The ratio of $600x$ to $125y$ $\rightarrow \dfrac{600x}{125y}$

$34p$ divided by 17 $\rightarrow \dfrac{34p}{17}$

2,500 per $1{,}000K$ $\rightarrow \dfrac{2{,}500}{1{,}000K}$

The quantity $(2x + 1)$ divided by 4 $\rightarrow \dfrac{2x+1}{4}$

EXERCISE 26.4

Represent the phrase by an algebraic expression.

1. The ratio of $200x$ to $25y$
2. $14p$ divided by 7
3. 1,500 per $10K$
4. The quantity $(2x - 3)$ divided by 5
5. The quotient of 100 and $2x$
6. The ratio of d to 100
7. 400 divided by $0.25x$
8. C divided by $2\pi r$
9. $7x$ over $2y$
10. The ratio of a to b
11. The quotient of $2x$ and 100
12. The quantity $(5x + 6)$ divided by 2
13. $8x$ divided by $\dfrac{1}{4}$
14. P divided by B
15. The quotient of 600 divided by t

Common Signal Words and Phrases for Equality

Common signal words and phrases that indicate equality are shown in the following table. Just as in arithmetic, the equal sign (=) is the symbol for equality.

Signal Word or Phrase	Example	Algebraic Symbolism
Equals, is equal to	$0.05n + 0.10d$ equals 48.85.	$0.05n + 0.10d = 48.85$
Is, are, will be	$(2K + 5) + (K + 5)$ will be 52.	$(2K + 5) + (K + 5) = 52$

Yields, gives	$2\%(10{,}000) + 3\%x$ yields 620.	$2\%(10{,}000) + 3\%x = 620$
Results in	$30\%x + 60\%(500)$ results in $40\%(x + 500)$.	$30\%x + 60\%(500) = 40\%(x + 500)$
Exceeds…by	78 exceeds $3n$ by 6.	$78 = 3n + 6$

Do not put periods at the end of algebraic symbolism for statements.

EXAMPLE

Represent the statement of equality by an equation.

$2w + 2(w + 5)$ yields 78. $\rightarrow 2w + 2(w + 5) = 78$

$(3n - 10)$ is equal to $(2n + 5)$. $\rightarrow (3n - 10) = (2n + 5)$

$15x + 6(300 - x)$ results in $9(300)$. $\rightarrow 15x + 6(300 - x) = 9(300)$

$\frac{d}{13.5}$ equals $\frac{20}{0.5}$. $\rightarrow \frac{d}{13.5} = \frac{20}{0.5}$

The signal words and phrases in this chapter are by no means all-inclusive. However, they are representative of the kinds of words and phrases that typically occur in word problems.

EXERCISE 26.5

Represent the statement of equality by an equation.

1. $2l + 2(l + 3)$ yields 52.
2. $(3x - 5)$ is equal to $(2x + 10)$.
3. $25x + 10(300 - x)$ results in 4,500.
4. $\frac{x}{12}$ equals $\frac{200}{3}$.
5. $6\%B$ yields 57.60.
6. $0.25q + 0.10(42 - q)$ is equal to 5.55.
7. $55t + 65t$ results in 624.
8. c^2 equals $8^2 + 15^2$.
9. $(l + 13)$ is $\frac{1}{2}P$.
10. n increased by 3 equals 15.
11. The sum of x and $\frac{1}{2}x$ is 63.
12. The product of w and $(w + 3)$ yields 70.
13. 95 exceeds $5y$ by 10.
14. $(n + 2)$ is 10 more than twice $(n + 1)$.
15. $\frac{1}{3}$ of x added to $\frac{1}{4}$ of x will be 35.

CHAPTER 27

Applying Algebra to Word Problems

Steps in Algebraic Problem-Solving

To be successful with solving word problems, you need to take a systematic approach. Here are suggested steps for using an algebraic approach to solving word problems.

Resist the urge to solve word problems by trial and error. That approach works only with simplistic problems and is often futile for more complicated situations.

Step 1. Understand the problem.

Is it a problem that fits a familiar type (for example, a number problem, percentage problem, mixture problem)? Identify what the problem is asking you to determine. In other words, what is the question? Look for words/phrases such as *determine, what is, how many, how far, how much, find,* and so on. Identify the unknown(s) that will lead to a solution.

Step 2. Represent the unknown(s) with variable expressions.

Choosing the first letter of the name of an unknown as its variable representation can help you keep variable names straight.

As you represent unknowns, make explicit statements such as "Let $x = \ldots$" so it is clear what each variable represents. Specify the variable's units, if applicable.

In problems with one or more unknowns, assign a variable name to one of the unknowns. Next, identify relationships that allow you to express the other unknowns in terms of that variable. If one of the unknowns is described in terms of another unknown, designate the variable as the unknown used in the description. For example, if Josie is twice as old as Micah, then let $M =$ Micah's age (in years) and $2M =$ Josie's age (in years).

Step 3. Analyze the question information.

Using the variable representations determined in Step 2, record what you know including units, if any. Write a statement of equality that accurately models the facts/relationships in the problem. If units are involved, check that the indicated calculations will result in proper units (see the next lesson for an additional discussion of this topic).

Step 4. Use algebraic symbolism to model the problem.

Represent the statement of equality from Step 3 by an equation. Reread the problem to make sure your equation accurately represents the situation given.

Step 5. Solve the equation.

Use algebra to solve the equation. For convenience you can omit units while solving equations, because you have already checked that the results will have the proper units.

Step 6. State your solution and assess its reasonableness.

State the solution in words. Did you answer the question asked? Did you include the units, if applicable? Does your solution make sense in the context of the problem?

Be flexible. The process is systematic, but not rigid. You can make impromptu modifications that fit your problem-solving style.

Keep in mind that problem solving seldom occurs in a linear fashion. Not infrequently, you will have to go back to a previous step. As you gain confidence in your problem-solving ability, you might skip steps, or even combine steps.

Nevertheless, the problem-solving process explained in this lesson can serve as a guide to assist you in solving a multitude of word problems.

When looking at the examples in this section, realize there are usually multiple ways to solve a problem. You might think of ways to reach the correct solutions other than the ones shown.

EXAMPLE

One number is 5 times a second number. The sum of the two numbers is 84. Find the numbers.

Represent the unknowns with variable expressions. Let $s =$ the smaller number. Then $5s =$ the larger number. Write a statement of equality.

s plus $5s$ is 84.

Write and solve an equation that represents your statement of equality.

Solve $s + 5s = 84$.

$$s + 5s = 84$$
$$6s = 84$$
$$\frac{6s}{6} = \frac{84}{6}$$
$$s = 14$$
$$5s = 70$$

State the solution in words. The two numbers are 14 and 70.

EXAMPLE

The sum of two integers is 17 and their product is 72. Find the smaller number.

Represent the unknowns with variable expressions. Let s = the smaller number. Then $17 - s$ = the larger number. Write a statement of equality.

s times $(17 - s)$ is 72.

Write and solve an equation that represents your statement of equality.

Solve $s(17 - s) = 72$.

$$s(17 - s) = 72$$
$$17s - s^2 = 72$$
$$0 = s^2 - 17s + 72$$
$$s^2 - 17s + 72 = 0$$
$$(s - 8)(s - 9) = 0$$
$$s - 8 = 0 \text{ or } s - 9 = 0$$
$$s = 8 \text{ or } s = 9 \text{ (reject, because } s \text{ is the smaller number)}$$

State the solution in words. The smaller of the two numbers is 8.

When you solve a quadratic equation, you can expect to obtain two values for the variable, both of which make the equation true. Most of the time in application problems, the question will provide information that will lead you to reject one of the two values obtained.

EXERCISE 27.1

For 1 to 4, represent all the unknowns in the statement with variable expressions.

1. The total number of nickels and dimes in a jar is 759.
2. There is 30 pounds in a mixture of candy that sells at \$11.50 per pound and candy that sells at \$19.90 per pound.
3. Nidhi's grandmother is 4 times as old as Nidhi.
4. Kat is 5 years younger than Richard.

Solve the next question.

5. One number is 12 more than twice another number. What are the numbers if their sum is 72?

Be Careful with Units

Word problems often involve quantities that are specified with units (such as 20 inches, 4.5 pounds, 2.5 hours, 5 years, 60 miles per hour, \$200, etc.). (See Appendix A for a Measurement Units and Conversions table.) In mathematical computations involving units, a completely defined quantity has both a numerical component and a units component. The units must undergo the same mathematical operations that are performed on the numerical component of the quantity. You perform operations on units like you do on numbers.

You can add or subtract units only if they can be expressed as like units.

EXAMPLE

- $20\text{ in} + 15\text{ in} = 35\text{ in}$
- $1\text{ hr} + 30\text{ min} = 1\text{ hr} + 0.5\text{ hr} = 1.5\text{ hr}$

You can multiply and divide units whether they are like or unlike. However, the resulting product or quotient must have meaning in the context of the problem.

Put units that follow the word *per* in a denominator.

EXAMPLE

- $(5\text{ ft})(8\text{ ft}) = 40\text{ ft}^2$
- $(60\text{ miles per hr})(2.5\text{ hr}) = \left(60\ \frac{\text{miles}}{\text{hr}}\right)(2.5\text{ hr}) = \left(60\ \frac{\text{miles}}{\cancel{\text{hr}}}\right)(2.5\ \cancel{\text{hr}})$
  $= 150\text{ miles}$

EXERCISE 27.2

Compute as indicated.

1. $300\text{ mL} + 200\text{ mL}$
2. $(50\text{ cm})(20\text{ cm})$
3. $85° - 10°$
4. $\frac{4}{3}\pi(6\text{ in})^3$
5. $\left(75\ \frac{\text{miles}}{\text{hr}}\right)(2\text{ hr})$
6. $\frac{13.5\text{ in}}{0.5\text{ in}}$
7. $\left(\frac{\$25}{\text{hr}}\right)(3.5\text{ hr})$
8. $2\%(\$1{,}400) + 1.5\%(\$2{,}000)$
9. $(10\text{ m})(6\text{ m})(4\text{ m})$
10. $4.5\text{ lb} \cdot \$15\text{ per pound}$

CHAPTER 28

Applications

Number Problems

In number problems, you are given information about one or more numbers. Your task is to find the value(s) of the number(s).

No units are involved in number problems.

EXAMPLE

▶ One number is 4 times another number. Twice the sum of the two numbers is 85. Find the larger number.

Represent the unknowns with variable expressions.

Let $s =$ the smaller number. Then $4s =$ the larger number.

Write a statement of equality.

2 times $(s + 4s)$ is 85.

Write and solve an equation that represents your statement of equality.

Solve $2(s + 4s) = 85$.

$$\begin{aligned} 2(s + 4s) &= 85 \\ 2(5s) &= 85 \\ 10s &= 85 \\ \frac{10s}{10} &= \frac{85}{10} \\ s &= 8.5 \end{aligned}$$

Find $4s$, the larger number.

$4s = 4(8.5) = 34$

State the solution in words: The larger of the two numbers is 34.

Consecutive integers follow each other in order and differ by 1.

EXAMPLE

▶ The greatest of three consecutive integers is 10 more than twice the second integer. What is the greatest of the three integers?

Represent the unknowns with variable expressions.

Let $n =$ the first integer, $n + 1 =$ the second integer, and $n + 2 =$ the third integer (the greatest one).

Write a statement of equality.

$(n + 2)$ is 10 more than 2 times $(n + 1)$.

Write and solve an equation that represents your statement of equality.

Solve $(n + 2) = 2(n + 1) + 10$.

$$\begin{aligned}
(n + 2) &= 2(n + 1) + 10 \\
n + 2 &= 2n + 2 + 10 \\
n + 2 &= 2n + 12 \\
n + 2 - 2n &= 2n + 12 - 2n \\
-n + 2 &= 12 \\
-n + 2 - 2 &= 12 - 2 \\
-n &= 10 \\
n &= -10
\end{aligned}$$

Find $n + 2$, the greatest integer.

$n + 2 = -10 + 2 = -8$

State the solution in words: The greatest of the three consecutive integers is -8.

Consecutive even integers and **consecutive odd integers** follow each other in order and differ by 2.

EXAMPLE

▶ The sum of the first and second of three consecutive odd integers is 35 less than 3 times the third integer. What are the three integers?

Represent the unknowns with variable expressions.

Let $n =$ the first odd integer, $n + 2 =$ the second odd integer, and $n + 4 =$ the third odd integer.

Write a statement of equality.

The sum of n and $n + 2$ is 35 less than three times $n + 4$.

Write and solve an equation that represents your statement of equality.

Solve $n + (n + 2) = 3(n + 4) - 35$.

$$\begin{aligned} n + (n+2) &= 3(n+4) - 35 \\ n + n + 2 &= 3n + 12 - 35 \\ 2n + 2 &= 3n - 23 \\ 2n + 2 - 2n &= 3n - 23 - 2n \\ 2 &= n - 23 \\ 2 + 23 &= n - 23 + 23 \\ 25 &= n \end{aligned}$$

Find $n + 2$ and $n + 4$.

$n + 2 = 25 + 2 = 27$

$n + 4 = 25 + 4 = 29$

State the solution in words: The three odd integers are 25, 27, and 29.

EXERCISE 28.1

Solve the following problems.

1. Find the greatest of three consecutive integers such that the sum of the greatest plus 5 times the least of the three integers is −250.

2. The sum of the first and 3 times the second of three consecutive even integers is 38 greater than twice the third integer. What are the three integers?

3. Two times a certain number is 6 less than 78 minus the same number. What is the number?

4. The square of a positive number exceeds the same number by 12. Find the number.

5. A number increased by 8 times the same number yields 189. What is the number?

6. Two-thirds of a number is 86. What is the number?

7. If a number is increased by 0.08 of itself, the result is 120.96. What is the number?

8. A number reduced by 25% of itself yields 195. What is the number?

9. A number divided by $\frac{1}{2}$ is 10. What is the number?

10. The quotient of a number and 0.25 equals 200 less than 6 times the same number.

Age Problems

In age problems, comparisons of ages are usually made in specified time periods (present, future, or past).

EXAMPLE

Josie is twice as old as Micah. Five years from now the sum of their ages will be 52. How old will Josie be in 10 years?

Represent the unknowns with variable expressions.

Let M = Micah's age now (in years). Then $2M$ = Josie's age now (in years).

Make a table to organize the information given.

When?	Micah's age (in years)	Josie's age (in years)	Sum (in years)
Now	M	$2M$	Not given
In 5 years	$M + 5$	$2M + 5$	52
In 10 years	$M + 10$	$2M + 10$	Not given

Write a statement of equality that expresses facts shown in the table.

$M + 5$ plus $2M + 5$ is 52.

Check units: years + years = years ✓

Write and solve an equation that represents your statement of equality.

Solve $(M + 5) + (2M + 5) = 52$.

$$\begin{aligned}
(M + 5) + (2M + 5) &= 52 \\
M + 5 + 2M + 5 &= 52 \\
3M + 10 &= 52 \\
3M + 10 - 10 &= 52 - 10 \\
3M &= 42 \\
\frac{3M}{3} &= \frac{42}{3} \\
M &= 14
\end{aligned}$$

Make sure you answer the question asked. In this age problem, after you obtain M, you must calculate $2M + 10$ to answer the question.

Find $(2M + 10)$, Josie's age 10 years from now.

$2M + 10 = 2(14) + 10 = 28 + 10 = 38$

State the solution in words: Josie will be 38 years old in 10 years.

EXERCISE 28.2

Solve the following problems.

1. Currently, Nidhi's grandmother is 4 times as old as Nidhi. Ten years ago, Nidhi's grandmother was 7 times as old as Nidhi. How old is Nidhi now?
2. Kat is 5 years younger than Richard. Ten years ago, Richard was twice Kat's age. How old is Kat now?
3. Currently, Pablo is 4 times as old as his son. In 16 years, he will be only twice his son's age. What are their ages now?
4. Currently, Hayley is one-fifth as old as her brother Nathan. Four years from now 3 times Hayley's age will equal Nathan's age. How old is Nathan now?
5. Monette is 6 years older than Juliet. In two years, Monette will be twice as old as Juliet. Find their present ages.
6. Kaxon is 12 years younger than Samuel. Three years ago, Samuel was 5 times as old as Kaxon. How old is Samuel now?
7. Loralei is 5 years older than Jonah. Four years ago, 8 times Jonah's age equaled 3 times Loralei's age. What is Loralei's present age?
8. Currently, Liam is 4 times as old as Henri. Six years ago, Liam was 10 times as old as Henri. What are their ages now?
9. Currently, the sum of the ages of Candice and her daughter Sophia is 45 years. Five years ago, Candice's age was 6 times Sophia's age. What is Sophia's age now?
10. The sum of the ages of Arbela and Loy is 48 years. In eight years, Arbela will be 3 times Loy's age. What are their ages now?

Ratio and Proportion Problems

A **ratio** is a quotient of two quantities. In word problems, if two quantities are in the ratio a to b and you know their sum is c, express the two quantities as ax and bx, where x is a common factor. Next, solve $ax + bx = c$, for x, and then compute ax or bx, whichever is needed.

EXAMPLE

The ratio of boys to girls in a classroom of 35 students is 3 to 4. How many girls are in the classroom?

Represent the unknowns with variable expressions.

Let $3x =$ the number of boys in the classroom and $4x =$ the number of girls in the classroom.

Write a statement of equality based on the following fact:

The number of boys plus the number of girls in the classroom is 35.

$3x$ plus $4x$ is 35.

Check units: No units are needed.

Write and solve an equation that represents your statement of equality.

Solve $3x + 4x = 35$

$$
\begin{aligned}
3x + 4x &= 35 \\
7x &= 35 \\
\frac{7x}{7} &= \frac{35}{7} \\
x &= 5
\end{aligned}
$$

Find $4x$, the number of girls.

$4x = 4(5) = 20$

State the solution in words: There are 20 girls in the classroom.

You can extend the strategy shown in this problem to three or more quantities.

A **proportion** is a mathematical statement that two ratios are equal. The statement $\frac{a}{b} = \frac{c}{d}$ is a proportion and is read "a is to b as c is to d." The **fundamental property of proportions** is that $\frac{a}{b} = \frac{c}{d}$ if and only if $ad = bc$. The numbers a, b, c, and d are the **terms** of the proportion. The products ad and bc are the **cross products** (illustrated below).

EXAMPLE

On a map, the distance between two cities is 13.5 inches. The scale on the map shows that 0.5 inches represents 20 miles. What is the distance, in miles, between the two cities?

Represent the unknown with a variable expression.

Let $d =$ the distance, in miles between the two cities.

Write a statement of equality between two ratios based on the facts given.

The ratio of d to 20 miles equals the ratio of 13.5 inches to 0.5 inches.

Check units: $\frac{\text{miles}}{\text{miles}} = \frac{\text{inches}}{\text{inches}}$ √ (because the units cancel out on both sides)

Write and solve an equation that represents your statement of equality.

Solve $\frac{d}{20 \text{ miles}} = \frac{13.5 \text{ in}}{0.5 \text{ in}}$, omitting units for convenience.

$$\frac{d}{20} = \frac{13.5}{0.5}$$
$$(d)(0.5) = (20)(13.5)$$
$$0.5d = 270$$
$$\frac{0.5d}{0.5} = \frac{270}{0.5}$$
$$d = 540$$

Solve proportions by setting cross products equal to each other.

State the solution in words: The distance between the two cities is 540 miles.

EXERCISE 28.3

Solve the following problems.

1. The ratio of women to men in a campus service organization of 54 students is 4 to 5. How many women are in the organization?
2. Raph earns $97 in 4 hours. At this rate, how many hours does he work to earn $485?
3. Kenzie saved $54 in 8 weeks. How long, in weeks, will it take her to save $243 at the same rate?
4. An RV can travel 270 miles on 18 gallons of gasoline. At this rate, how many miles can the RV travel on 24 gallons of gasoline?
5. Two partners divide their profits for the month in the ratio of 3 to 4. How much will each get in January if the profit for that month is $3,500?

6. A tree casts a shadow of 30 feet, while a 6-foot pole nearby casts a shadow of 5 feet. What is the height, in feet, of the tree?

7. The ratio of math teachers to English teachers attending the conference is 2 to 3. How many math teachers are in attendance if there are a total of 375 math and English teachers at the conference?

8. A homeowner pays $760 taxes on land property assessed at $38,000. What will be the taxes on a land property assessed at $63,350 if the same rate is used?

9. Baylee drives 304 miles in 4 hours without stopping. At this speed, how long, in hours, would it take Baylee to drive 190 miles without stopping?

10. A 4-by-6-inch picture is enlarged, so that the longest side is 15 inches. What is the width, in inches, of the enlarged picture?

Mixture Problems

In a mixture problem, the amount (or value) of a substance before mixing equals the amount (or value) of that substance after mixing.

EXAMPLE

How many milliliters of a 30% alcohol solution must be added to 500 milliliters of a 60% alcohol solution to yield a 40% alcohol solution?

Represent the unknowns with variable expressions.

Let $x =$ the amount (in milliliters) of the 30% alcohol solution that must be added. Then $x + 500 =$ the amount (in milliliters) in the final solution.

Make a table to organize the information given.

When?	Percent alcohol strength	Amount (in milliliters)	Amount of alcohol (in milliliters)
Before	30%	x	$30\%x$
	60%	500	$60\%(500)$
After	40%	$x + 500$	$40\%(x + 500)$

Using the table, write a statement of equality.

$30\%x$ plus $60\%(500)$ is $40\%(x + 500)$.

Check units: milliliters + milliliters = milliliters ✓

Write and solve an equation that represents your statement of equality.

Solve $30\%x + 60\%(500) = 40\%(x + 500)$.

$$
\begin{aligned}
30\%x + 60\%(500) &= 40\%(x + 500) \\
0.30x + 0.60(500) &= 0.40(x + 500) \\
0.30x + 300 &= 0.40x + 200 \\
0.30x + 300 - 0.40x &= 0.40x + 200 - 0.40x \\
-0.10x + 300 &= 200 \\
-0.10x + 300 - 300 &= 200 - 300 \\
-0.10x &= -100 \\
\frac{-0.10x}{-0.10} &= \frac{-100}{-0.10} \\
x &= 1000
\end{aligned}
$$

State the solution in words: 1000 milliliters of the 30% alcohol solution must be added.

EXERCISE 28.4

Solve the following problems.

1. How many milliliters of distilled water must be added to 1000 milliliters of a 70% alcohol solution to yield a 50% alcohol solution?
2. A candy store owner mixes candy that normally sells for \$11.50 per pound and candy that normally sells for \$19.90 per pound to make a 30-pound mixture to sell at \$17.10 per pound. To make sure that \$17.10 per pound is a fair price, how many pounds of the \$11.50 candy should the owner use in the mixture?
3. How many ounces of pure vinegar must be added to 80 ounces of a 10% solution to make a 25% solution?
4. How many milliliters of a 10% nitric acid solution must be added to 1400 milliliters of a 25% nitric acid solution to make a 20% nitric acid solution?
5. How many quarts of 100% antifreeze must be added 10 quarts of a 20% antifreeze solution to make a 50% antifreeze solution?
6. A dairy scientist wants to make a milk mixture that contains 4% butterfat. How many quarts of milk containing no butterfat must be added to 400 quarts of milk containing 5% butterfat to make a milk mixture that contains 4% butterfat?
7. A grocer mixes nuts worth \$10 per pound with nuts worth \$15 per pound to make a mixture of 90 pounds of nuts to sell at \$12 per pound. To make sure that \$12 per pound is a fair price, how many pounds of each should the grocer use in the mixture?
8. A coffee shop manager mixes coffee worth \$8 per pound with 20 pounds of coffee worth \$14 per pound to get a mixture that will be sold for \$10 per pound. To make sure that \$10 per pound is a fair price, how many pounds of the cheaper coffee should the manager use in the mixture?
9. How many liters of a 4% hydrochloric acid solution must be added to a 20% hydrochloric acid solution to obtain 10 liters of a 12% hydrochloric acid solution?
10. A butcher mixes ground beef that is 80% lean with ground beef that is 88% lean to make 200 pounds of a ground beef mixture that is 83% lean. How many pounds of each should the butcher use?

Of course, in coin problems, you must assume there are no rare coins in the collections that would be worth more than their face values.

Coin Problems

In a coin problem, the value of a collection of coins equals the sum of the values of the coins in the collection.

EXAMPLE

Jude has a jar containing 759 U. S. nickels and dimes that have a total value of \$53.30. How many nickels are in the jar?

Represent the unknowns with variable expressions.

Let $n =$ the number of nickels. Then $759 - n =$ the number of dimes in the jar.

Make a table to organize the information given.

Denomination	Nickels	Dimes	Total
Face value per coin (in dollars)	0.05	0.10	N/A
Number of coins	n	$759 - n$	759
Value of coins (in dollars)	$0.05n$	$0.10(759 - n)$	53.30

Using the table, write a statement of equality.

$0.05n$ plus $0.10(759 - n)$ equals 53.30.

Check units: $\$ + \$ = \$$ ✓

Write and solve an equation that represents your statement of equality.

Solve $0.05n + 0.10(759 - n) = 53.30$.

$$\begin{aligned}
0.05n + 0.10(759 - n) &= 53.30 \\
0.05n + 75.90 - 0.10n &= 53.30 \\
-0.05n + 75.90 &= 53.30 \\
-0.05n + 75.90 - 75.90 &= 53.30 - 75.90 \\
-0.05n &= -22.60 \\
\frac{-0.05n}{-0.05} &= \frac{-22.60}{-0.05} \\
n &= 452
\end{aligned}$$

State the solution in words: There are 452 nickels in the jar.

EXERCISE 28.5

Solve the following problems.

1. Nashi has a collection of change consisting of 200 nickels and dimes. The coins have a total value of $13.50. How many dimes are in the collection?
2. Ennis has $3.10 in nickels and dimes. He has 14 more nickels than dimes. Find the number of each.
3. Ronin has $250 in denominations of $5 and $10 bills only. He has 3 times as many $5 bills as he has $10 bills. How many of each does he have?
4. Latsha has 4 times as many $5 bills as $1 bills, with the total amounting to $84. How many bills of each kind does she have?
5. Jermo received 2 more dimes than quarters in exchange for a $10 bill. How many dimes did he receive?
6. Barb has $2.34 in dimes, nickels, and pennies. She has 3 times as many dimes as nickels and 6 more pennies than dimes. How many of each kind of coin does she have?
7. Joyce has 16 coins consisting of quarters and nickels. The total value of the coins is $1.40. How many quarters does she have?
8. Willow has $4.30 in nickels and dimes, totally 52 coins. How many nickels does Willow have?
9. Cael has $7.50 consisting of quarters and nickels. If the number of nickels is 6 more than the number of quarters, how many nickels and how many quarters does Cael have?
10. Scarlett has 45 coins consisting of nickels, dimes, and quarters, for a total value of $7.00. The number of dimes exceeds the number of nickels by 5. How many quarters does Scarlett have?

Rate-Time-Distance Problems

For the formula $d = rt$, the time units for the rate must match the time period units.

The distance formula is $d = rt$, where d is the distance a vehicle travels at a uniform rate of speed, r, for a given length of time, t.

EXAMPLE

A car and a van leave the same location at the same time. The car travels due east at 70 miles per hour. The van travels due west at 60 miles per hour. How long will it take for the two vehicles to be 325 miles apart?

Represent the unknown with a variable expression.

Let $t =$ the time in hours the two vehicles will be 325 miles apart.

Make a table to organize the information given.

Vehicle	Rate (in miles per hour)	Time (in hours)	Distance (in miles)
Car	70	t	$70t$
Van	60	t	$60t$

Sketch a diagram.

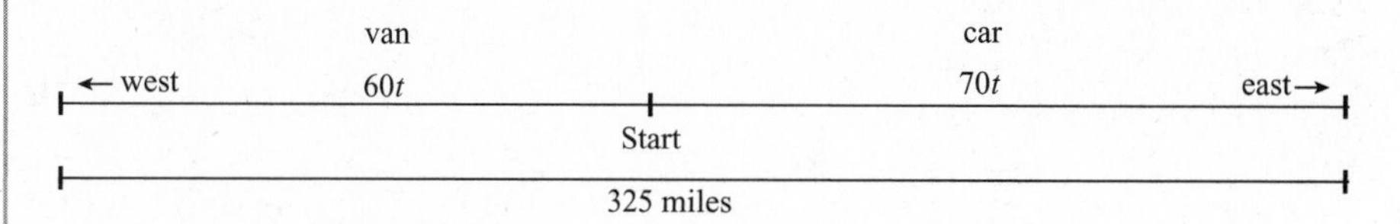

The diagram shows that the sum of the distances traveled by the two vehicles equals 325 miles.

Write a statement of equality that expresses the facts shown: $60t$ plus $70t$ is 325 miles.

Check units: $\left(\frac{\text{miles}}{\cancel{\text{hr}}}\right)(\cancel{\text{hr}}) + \left(\frac{\text{miles}}{\cancel{\text{hr}}}\right)(\cancel{\text{hr}}) = \text{miles} + \text{miles} = \text{miles}$ ✓.

Write and solve an equation that represents your statement of equality.

Solve $60t + 70t = 325$.

$$60t + 70t = 325$$
$$130t = 325$$
$$\frac{130t}{130} = \frac{325}{130}$$
$$t = 2.5$$

State the solution in words: It will take 2.5 hours for the two vehicles to be 325 miles apart.

EXERCISE 28.6

Solve the following problems.

1. One vehicle, traveling at an average speed of 70 miles per hour, leaves city A on the way to city B, a distance of 270 miles. At the same time, a second vehicle, traveling at an average speed of 65 miles per hour leaves city B on the way to city A. If both vehicles maintain their respective average speeds, in how many hours will the two vehicles pass each other?
2. At 8 p.m., a car and van leave the same location. The car travels due east at 70 miles per hour. The van travels due west at 60 miles per hour. If both vehicles continue to travel as stated, at what clock time will the two vehicles be 325 miles apart?
3. A car and a truck are 540 miles apart. The two vehicles start driving toward each other at exactly the same time. The car travels at a speed of 65 miles per hour and the truck travels at a speed of 55 miles per hour. How soon, in hours, will the two vehicles arrive at the same location if both continue at their given speeds without making any stops?
4. What average speed, in miles per hour, did a car travel to overtake a bus in 3 hours if the bus left 1 hour before the car, traveling at an average speed of 60 miles per hour?
5. Two trains leave a station traveling in opposite directions, one at an average speed of 55 miles per hour and the other at an average speed of 50 miles per hour. In how many hours will they be 315 miles apart?
6. How many minutes will it take an airplane flying at an average speed of 550 miles per hour to cover a distance of 137.5 miles?
7. Mora and Leith leave on their bicycles from the same place, but ride in opposite directions. Mora rides twice as fast as Leith, and in 4 hours, they are 24 miles apart. What is Mora's average speed in miles per hour?
8. Two bicycle riders start at the same time from opposite ends of a 45-mile-long trail. One rider travels at an average speed of 16 miles per hour and the other rider travels at an average speed of 14 miles per hour. In how many hours after they begin will they meet each other?
9. A river has a current of 3 miles per hour. A boat travels downstream in the river for 3 hours with the current, and then returns upstream the same distance against the current in 4 hours. What is the boat's speed, in miles per hour, when there is no current?
10. A car and a bus leave the same location at the same time headed in the same direction. The average speed of the car is 30 miles per hour slower than twice the speed of the bus. In 2 hours, the car is 20 miles ahead of the bus. What is the car's average speed, in miles per hour?

Work Problems

In work problems, the portion of a task completed in a unit time period is the reciprocal of the amount of time it takes to complete the task. For example, if it takes Myla four hours, working alone, to paint a hallway, then the portion of the hallway she can paint in one hour, working alone, is $\frac{1}{4}$ of the hallway.

The unit time for the work done must be the same for all, individually and combined.

Importantly, when two or more individuals (machines, devices, entities, etc.) work together, the portion of the work done per unit time by the first individual plus the portion of the work done per unit time by the second individual plus the portion of the work done per unit time by the third individual and so on equals the portion of the work done per unit time when all individuals work together.

EXAMPLE

Myla can paint a hallway in 4 hours working alone. Kenton can do the same task in 6 hours working alone. How long (in hours) will it take Myla and Kenton, working together, to paint the hallway?

Represent the unknown with a variable expression.

Let $t =$ the time it will take Myla and Kenton, working together, to paint the hallway.

Make a table to organize the information given.

Situation	Time (in Hours)	Portion of Hallway per Hour
Myla working alone	4	$\frac{1}{4}$
Kenton working alone	6	$\frac{1}{6}$
Myla and Kenton working together	t	$\frac{1}{t}$

Using the table, write a statement of equality based on the following fact:

The portion of the hallway done per hour by Myla plus the portion of the hallway done per hour by Kenton equals the portion of the hallway done per hour by Myla and Kenton working together.

$\frac{1}{4}$ of the hallway per hour plus $\frac{1}{6}$ of the hallway per hour equals $\frac{1}{t}$ of the hallway per hour.

Check units: $\frac{\text{hallway}}{\text{hour}} + \frac{\text{hallway}}{\text{hour}} = \frac{\text{hallway}}{\text{hour}}$. ✓

Write and solve an equation that represents your statement of equality.

Solve $\frac{1}{4} + \frac{1}{6} = \frac{1}{t}$.

$$\frac{1}{4} + \frac{1}{6} = \frac{1}{t}$$
$$\frac{3}{12} + \frac{2}{12} = \frac{1}{t}$$
$$\frac{5}{12} = \frac{1}{t}$$
$$\frac{\cancel{12}t}{1} \cdot \frac{5}{\cancel{12}} = \frac{12\cancel{t}}{1} \cdot \frac{1}{\cancel{t}}$$
$$\frac{\cancel{5}t}{\cancel{5}} = \frac{12}{5}$$
$$t = 2.4$$

State the solution in words: The time it will take Myla and Kenton, working together, to paint the hallway is 2.4 hours.

The time it will take when two or more individuals (or machines, devices, entities, etc.) work together (in a positive way) is always less than the least individual time.

EXERCISE 28.7

Solve the following problems.

1. A water tank can be filled by one pipe by itself in 5 hours and by a second pipe by itself in 3 hours. How many hours will it take the two pipes together to fill the tank?
2. Working alone, Loria can mow a lawn in 24 minutes, and Darius can do it in 48 minutes. How long, in minutes, does it take the two of them working together to mow the lawn?
3. Kelsey can mow a large field twice as quickly as Imogene. Together they can do it in 2 hours. How long, in hours, would it take for each girl to mow the field alone?
4. It takes 40 minutes for one sprinkler to water a lawn, and 1 hour for a smaller sprinkler to do it. How long, in minutes, will it take to water the lawn if both sprinklers operate at the same time?
5. A tank can be filled with water in 16 hours. It takes 20 hours to drain the tank. If the tank is empty, how many hours will it take to fill the tank if the drain is open while the tank is being filled?
6. It takes Jakee 1 hour and 40 minutes to wash the family car, but it takes her younger sister 2 hours 30 minutes to do it. Working together, how many minutes will it take the two sisters to wash the family car?
7. Machine A can produce 500 items in 3 hours. Machine B can do it

3 times as fast. If both machines operate at the same time, how long, in hours, will it take them to produce 500 items?

8. A machine can do a job in 9 hours, and a second machine can do it in 18 hours. After the first machine has operated for 3 hours, the second machine is put into operation and together they complete the job. How many total hours did it take to complete the job?

9. Madison can paint a room in 4 hours, but she and her roommate do it together in 3 hours. What portion of the room does the roommate paint per hour?

10. A tank can be filled by one pipe in 8 hours and can be emptied by another pipe in 12 hours. If the tank is empty, how long, in hours, will it take to fill the tank if both pipes are open?

In word problems, a percent without a base is usually meaningless. Be sure to identify the base associated with each percent mentioned in a problem.

Percentage Problems

In simple percentage problems use the formula $P = RB$, where P is the percentage (the "part of the whole"), R is the rate (the quantity with a % sign or the word *percent* attached), and B is the base (the "whole amount").

EXAMPLE

Sasha works at an electronics store that pays sales personnel a commission rate of 2% on total sales. Last week, Sasha's sales totaled $2,812. What commission did Sasha earn last week?

Represent the unknown with a variable expression.

Let $P =$ the amount of Sasha's commission (in dollars).

Write a statement of equality using the formula $P = RB$ and the following facts:

Sasha's commission is P; R is 2%; and B is $2,812.

P equals 2% of $2,812.

Check units: $ = $ ✓ There are no units associated with a percent.

Write and solve an equation that represents your statement of equality.

Solve $P = (2\%)(2{,}812)$.

$P = (2\%)(2{,}812)$
$P = (0.02)(2{,}812)$
$P = 56.24$

Change percents to decimals or fractions to perform calculations.

State the solution in words: Sasha earned $56.24 in commission last week.

EXAMPLE

An online store offered a 20% discount on all clothing items during a 2-day sale. Ricardo got \$30.80 off the price of a coat he purchased during the sale. What was the original price of the coat?

Represent the unknown with a variable expression.

Let $B =$ the original price of the coat (in dollars).

Write a statement of equality using the formula $P = RB$ and the following facts:

The original price of the coat is B; R is 20%; and P is \$30.80.

\$30.80 is 20% of B.

Check units: \$ = \$ √

Write and solve an equation that represents your statement of equality.

Solve $38.30 = 20\%B$.

$$30.80 = 20\%B$$
$$30.80 = 0.20B$$
$$\frac{30.80}{0.20} = \frac{0.20B}{0.20}$$
$$154 = B$$

State the solution in words: The original price of the jacket was \$154.

EXAMPLE

Sophia pays a sales tax of \$9.90 on a camera that cost \$120. What is the sales tax rate for the purchase?

Represent the unknown with a variable expression.

Let $R =$ the sales tax rate for the purchase.

Write a statement of equality using the formula $P = RB$ and the following facts:

The sales tax rate is R; P is \$9.90; and B is \$120.

\$9.90 is R times \$120.

Check units: \$ = \$ √

$9.90 = R(120)$

$$9.90 = R(120)$$
$$\frac{9.90}{120} = \frac{R(120)}{120}$$
$$0.0825 = R$$

Write the answer as a percent: $R = 0.0825 = 8.25\%$

State the solution in words: The sales tax rate for the purchase is 8.25%.

EXERCISE 28.8

Solve the following problems.

1. An online store specializes in high-security luggage. New customers get a discount of 15% on the first order. Before sales tax, how much money is saved when a new customer makes a first order of a high-security, soft-sided, durable, wheeled luggage priced at $295?

2. Ash works as a sales clerk at an electronics store that pays sales personnel a commission rate of 3% on total sales. What were Ash's total sales last week if she earned $55.35 in commission?

3. A customer saved $1,624 on a dining room set that had an original price of $5,800. The amount saved is what percent of the original price?

4. Sage paid $216 for a watch. If the watch was discounted 25%, what was the regular price of the watch?

5. A dealer sold a television for $650, yielding a profit of 30% on the cost to the dealer. How much did the dealer pay for the television?

6. Jaylynn works as a salesperson at a computer store. She receives $600 per week as salary and an additional 4% commission on sales. If Jaylynn received $1,450 as total salary, what was the amount of her sales for that week?

7. A number increased by 4% is 52. What is the number?

8. A price marked up by 250% is $210. What is the original price?

9. What is the sales tax rate if a customer pays $7.84 sales tax on a $98 pair of shoes?

10. What is 109.3% of $620?

In the formula $I = Prt$, the interest rate time units must match the time period units.

Simple Interest Problems

The simple interest formula is $I = Prt$, where I is the simple interest accumulated on a principal, P, at a simple interest rate, r, per time period for t time periods.

EXAMPLE

How many years will it take \$10,000 invested at 2% annual interest to earn \$1,400 in interest? Note: 2% annual interest means a simple interest rate of 2% per year.

Represent the unknown with a variable expression.

Let $t =$ the time (in years) it will take the investment to earn \$1,400 in interest at the given rate.

Write a statement of equality using the formula $I = Prt$ and the following facts:

The time period is t; I is \$1,400; P is \$10,000; and r is 2% per year.

\$1,400 is \$10,000 times 2% per year times t (in years).

Check units: $\$ = (\$)\left(\frac{2\%}{\cancel{yr}}\right)(\cancel{yr})$

Write and solve an equation that represents your statement of equality.

Solve $1,400 = (10,000)2\%(t)$.

$$\begin{aligned} 1,400 &= (10,000)(2\%)(t) \\ 1,400 &= 200t \\ \frac{1,400}{200} &= \frac{200t}{200} \\ 7 &= t \end{aligned}$$

State the solution in words: It will take 7 years for an investment of \$10,000 to earn \$1,400 in interest at a rate of 2% per year.

EXERCISE 28.9

Solve the following problems.

1. How much interest is earned on \$15,000 invested at 1.5% annual interest for 8 years?
2. How many years will it take \$5,000 invested at 2% annual interest to earn \$400 in interest?
3. A teacher invests \$4,800 at a simple interest rate of 2% per year. How much interest does the investment earn in 3 years?
4. What is the simple interest rate per year if an investment of \$3,500 earns \$262.50 in 5 years?
5. An investment at an annual rate of 4% earns \$34 less than an investment at an annual rate of 3%. Find the amounts of the two investments if the total amount invested is \$9,300.

6. If the interest on a principal invested at a simple interest rate of 2% per year for 2 years is \$156, what is the principal?

7. Ari invests a certain amount of money at 2% annual interest and twice as much at 4%. How much is invested at each rate if the total annual interest from both investments is \$150?

8. Izy invests \$7,000, some of it at 2% annual interest and the rest at 3% annual interest. How much money is invested at each rate if the amount of annual interest on each investment is the same?

9. A federal credit union is offering 2.5% simple interest on a certificate of deposit. How much interest will a \$2,500 certificate of deposit earn at the end of one year?

10. Ace borrowed \$2,000 from a friend and agreed to pay 12% annual interest for a period of 3 years. At the end of 3 years, how much principal plus interest will Ace owe his friend?

Geometry Problems

In problems about geometric figures, it is usually helpful to make a sketch to assist you in visualizing the question information. (See Appendix B for a list of familiar geometry formulas.)

EXAMPLE

▶ The length of a rectangular lawn is 3 feet longer than its width. The lawn's area is 70 ft^2. What is the lawn's length, in feet?

Represent the unknowns with variable expressions.

Let $w =$ the lawn's width (in feet). Then $w + 3 =$ the lawn's length (in feet).

Make a sketch to show the question information.

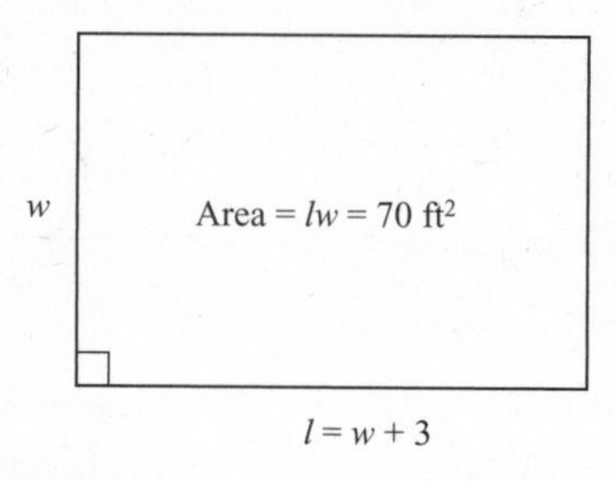

Write a statement of equality.

70 ft^2 equals $w + 3$ times w.

Check units: ft^2 = (ft)(ft) = ft^2 ✓

Write and solve an equation that represents your statement of equality.

Solve $70 = (w + 3)(w)$.

$$\begin{aligned} 70 &= (w + 3)(w) \\ 70 &= w^2 + 3w \\ w^2 + 3w - 70 &= 0 \\ (w + 10)(w - 7) &= 0 \\ w + 10 &= 0 \text{ or } w - 7 = 0 \\ w &= -10 \text{ (reject) or } w = 7 \end{aligned}$$

Reject negative values for dimensions of geometric figures because dimensions are always nonnegative.

Find $w + 3$, the length of the lawn.

$w + 3 = 7 + 3 = 10$

State the solution in words: The length of the lawn is 10 feet.

EXERCISE 28.10

Solve the following problems.

1. The area enclosed by a rectangular fence is 162 m^2. The length of the fence is twice its width. What are the fence's dimensions, in meters?

2. A rectangular flower box is 36 inches long, 6 inches high, and has a volume of 1,728 in^3. What is its width?

3. The perimeter of a rectangular field is 1,700 feet. If the length of the field is 500 feet, what is the area of the field in feet^2?

4. The area of a triangle is 108 inches^2. Find the length, in inches, of the triangle's base if the altitude has a measure of 18 inches.

5. What is the approximate area, in inches^2, of the cross section of a tree truck that has a circumference of 20π inches ? (Use $\pi \approx 3.14$.)

6. The measures of two angles of a triangle are 42° and 63°. What is the measure of the third angle?

7. In a right triangle, the hypotenuse has a length of 34 centimeters, and the length of one of the legs is 16 centimeters. What is the length of the other leg of the right triangle?

8. The length of a rectangular play area is 7 feet more than its width. What is the width of the play area if its area is 60 feet^2?

9. What is the diameter of a circle whose area is 64π meter^2?

10. If the perimeter of a rectangle is 36 centimeters, and one side is 2 centimeters shorter than the other, what are the rectangle's dimensions?

CHAPTER 29

Introduction to Functions

When discussing functions, the terms *x-coordinate, input value* or *input,* and *x-value* are interchangeable. Similarly, the terms *y-coordinate, output value* or *output,* and *y-value* are interchangeable.

Defining a Function

A **function** is a set of ordered pairs (x, y) in which each x-coordinate (**input value** or x-value) is paired with one and only one y-coordinate (**output value** or y-value). Thus, in a function no two ordered pairs have the same x-coordinate but different y-coordinates. Often, single letters such as f, g, and h designate functions.

EXAMPLE

- $f = \{(-5,5),(-2,5),(0,5),(2,5),(5,5)\}$ → is a function.

In a function, y-values of different ordered pairs can be the same.

- $w = \{(-3,4),(5,1),(4,3),(6,3),(4,2)\}$ → is not a function because (4, 3) and (4, 2) have the same x-coordinate.

In a function, x-values of different ordered pairs are never the same.

- $g = \{(-3,9),(-2,4),(-1,1),(0,0),(1,1),(2,4),(3,9)\}$ → is a function.
- $h = \left\{\left(-2,\frac{1}{4}\right),\left(-1,\frac{1}{2}\right),(0,1),(1,2),(2,4),(3,8)\right\}$ → is a function.

The **domain** of a function is the set of all x-coordinates of the ordered pairs in the function. The **range** of a function is the set of all y-coordinates of the ordered pairs in the function.

EXAMPLE

$f=\{(-5,5),(-2,5),(0,5),(2,5),(5,5)\}$ → Domain: $\{-5,-2,0,2,5\}$; Range: $\{5\}$

A y-value that appears multiple times in a function is listed only once in the range.

$t=\{(-1,-5),(0,-3),(1,-1)(2,1),(3,3)\}$ → Domain: $\{-1,0,1,2,3\}$; Range: $\{-5,-3,-1,1,3\}$

$g=\{(-3,9),(-2,4),(-1,1),(0,0),(1,1),(2,4),(3,9)\}$ → Domain: $\{-3,-2,-1,0,1,2,3\}$; Range: $\{0,1,4,9\}$

$h=\left\{\left(-2,\frac{1}{4}\right),\left(-1,\frac{1}{2}\right),(0,1),(1,2),(2,4),(3,8)\right\}$ → Domain: $\{-2,-1,0,1,2,3\}$; Range: $\left\{\frac{1}{4},\frac{1}{2},1,2,4,8\right\}$

EXERCISE 29.1

For 1 to 10, state Yes or No as to whether the set of ordered pairs is a function.

1. $f=\{(-5,-5),(-2,-2),(0,0),(2,2),(5,5)\}$
2. $f=\{(9,-3),(1,-1),(0,0)(1,1),(9,3)\}$
3. $g=\left\{\begin{matrix}(-3,8)(-2,3),(-1,0),(0,-1),(1,0),(2,3),\\(3,8)\end{matrix}\right\}$
4. $g=\{(-3,5)(-2,0),(-1,-3),(0,-4),(1,-3)\}$
5. $h=\{(-4,29)(2,2),(5,-11.5),(9,-29.5)\}$
6. $h=\{(1,5),(2,10),(3,15),(4,20),\ldots\}$
7. $f=\left\{\begin{matrix}(-8,30),(0,-6),(2,-15),(4,-24),\\(10,-51)\end{matrix}\right\}$
8. $f=\{(1,4),(2,8),(3,12),(4,16),(5,15),\ldots\}$
9. $g=\left\{\left(\frac{1}{2},-5\right)\right\}$
10. $g=\{(2,2),(4,-24),(5,-10),(5,0)\}$

For 11 to 20, state the domain and range of the given function.

11. $f = \{(-5,-5),(-2,-2),(0,0),(2,2),(5,5)\}$

12. $f = \left\{\left(-\frac{1}{2},0\right),\left(\frac{1}{2},0\right)\right\}$

13. $g = \left\{\begin{matrix}(-3,8),(-2,3),(-1,0),(0,-1),(1,0),(2,3),\\(3,8)\end{matrix}\right\}$

14. $g = \{(-3,5),(-2,0),(-1,-3),(0,-4),(1,-3)\}$

15. $h = \{(-4,29),(2,2),(5,-11.5),(9,-29.5)\}$

16. $h = \{(1,5),(2,10),(3,15),(4,20),\ldots\}$

17. $f = \left\{\begin{matrix}(-8,30),(0,-6),(2,-15),(4,-24),\\(10,-51)\end{matrix}\right\}$

18. $f = \{(1,4),(2,8),(3,12)(4,16),(5,20),\ldots\}$

19. $g = \left\{\left(\frac{1}{2},-5\right)\right\}$

20. $f = \{(1,3.5),(2,3.0),(4,2.2),(6,1.4),(9,1.0)\}$

Evaluating Functions

Evaluating a function means determining an output value that corresponds to a given input value. In an equation that defines a function f, the **function notation** $f(x)$ replaces y, and the equation is written in the form: $f(x) =$ an expression that contains only the x variable. The notation $f(x)$ is read "f of x." It denotes the output value y when x is the input value. When functions are defined by equations, refer to functions by their defining equations (e.g., say, "the function $f(x) = 2x - 3$)".

$f(x)$ does not mean f times x. It simply denotes the output that corresponds to the input x.

EXAMPLE

- Given $f(x) = 2x - 3$, then $f(-1) = 2(-1) - 3 = -2 - 3 = -5$.

To avoid careless errors, enclose substituted values in parentheses, when needed.

- Given $g(x) = x^2$, then $g(3) = (3)^2 = 9$.
- Given $h(x) = 2^x$, then $h(-2) = 2^{-2} = \frac{1}{2^2} = \frac{1}{4}$.
- Given $r(x) = \frac{1}{x-2}$, then $r(0) = \frac{1}{(0)-2} = \frac{1}{-2} = -\frac{1}{2}$.
- Given $t(x) = \sqrt{4-x}$, then $t(-5) = \sqrt{4-(-5)}$.

$$= \sqrt{4+5}$$
$$= \sqrt{9} = 3$$

Mathematicians use R to represent the set of real numbers.

When a function is defined by an equation, the domain is the set of all possible input values from R, the set of real numbers, and the range is the set of all possible output values from R. If no domain is specified, then it's understood that the domain is R, except for values that must be excluded. An **excluded value** is a value for the input value that would yield a value for the output that is undefined over R. Routinely, values that lead to division by zero or to square roots (or even roots) of negative numbers are excluded.

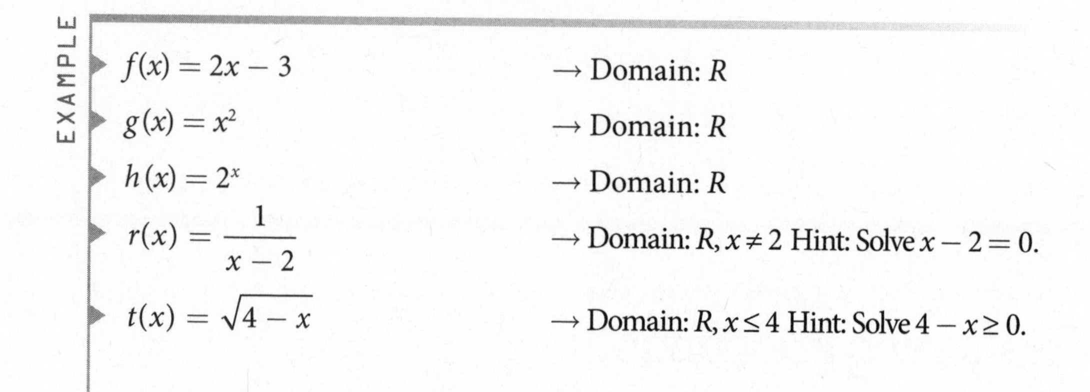

EXERCISE 29.2

For 1 to 3, evaluate $f(x) = \frac{1}{2}x + 5$ as indicated.

1. Find $f(-2)$.
2. Find $f(0)$.
3. Find $f\left(\frac{1}{2}\right)$.

For 4 to 6, evaluate $g(x) = 2x^2 - 5x - 3$ as indicated.

4. Find $g(-2)$.
5. Find $g(0)$.
6. Find $g(3)$.

For 7 to 9, evaluate $h(x) = 100(2^{x-1})$ as indicated.

7. Find $h(-1)$.
8. Find $h(0)$.
9. Find $h(5)$.

For 10 to 12, evaluate $t(x) = \sqrt{2x-1}$ as indicated.

10. Find $t(2)$.
11. Find $t(3)$.
12. Find $t(5)$.

For 13 to 15, evaluate $r(x) = \dfrac{5}{2x+1}$ as indicated.

13. Find $r(-3)$.

14. Find $r(0)$.

15. Find $r(2)$.

For 16 to 20, state the domain of the given function.

16. $f(x) = -3x + 5$

17. $g(x) = \dfrac{5x}{2x-6}$

18. $h(x) = 3^{x-2}$

19. $r(x) = x^2 - x - 12$

20. $t(x) = \sqrt{2x-1}$

CHAPTER 30

Graphs of Functions

Vertical Line Test

When a function f is defined by an equation, the graph f is the graph of the equation $y = f(x)$. Graphs of functions always pass the **Vertical Line Test**, meaning no vertical line crosses the graph more than once. Use the Vertical Line Test to visually determine whether a graph is the graph of a function.

The graphs of nonvertical lines are graphs of functions because every vertical line crosses a nonvertical line only once.

EXAMPLE

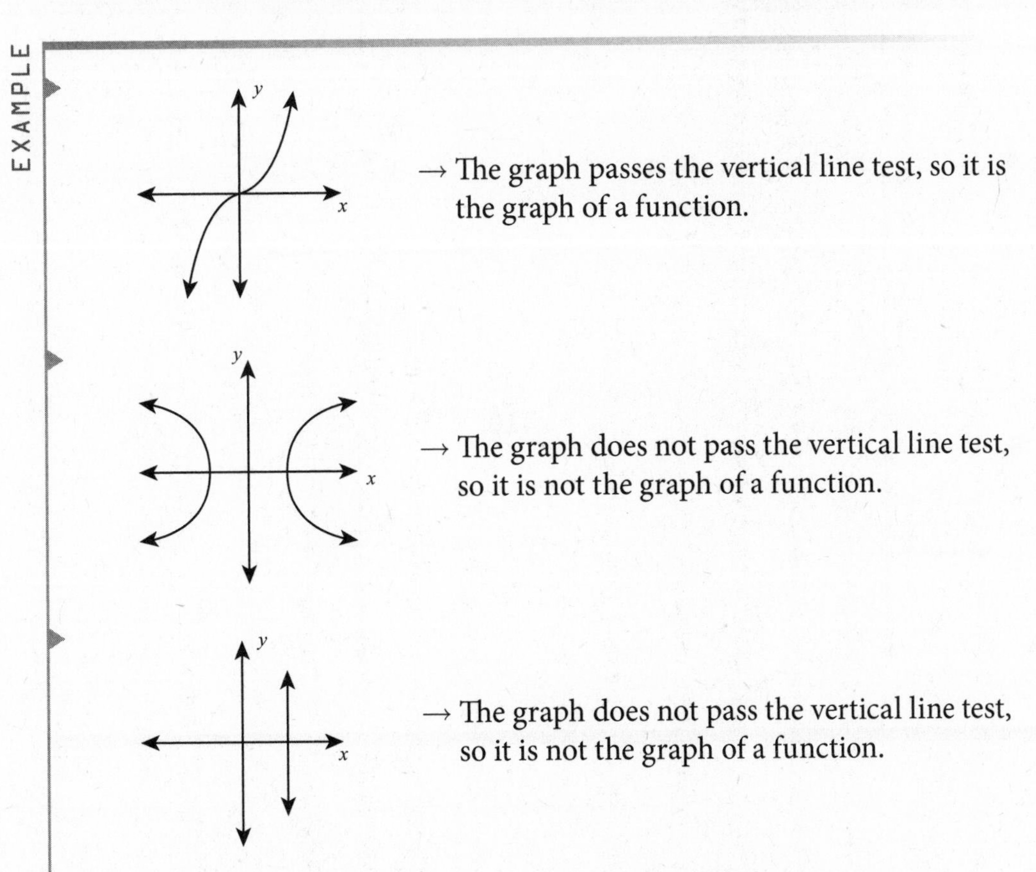

EXERCISE 30.1

State Yes or No as to whether the given graph represents a function.

1.

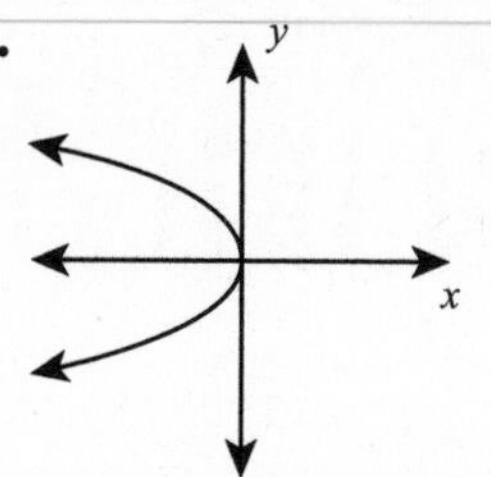

2.

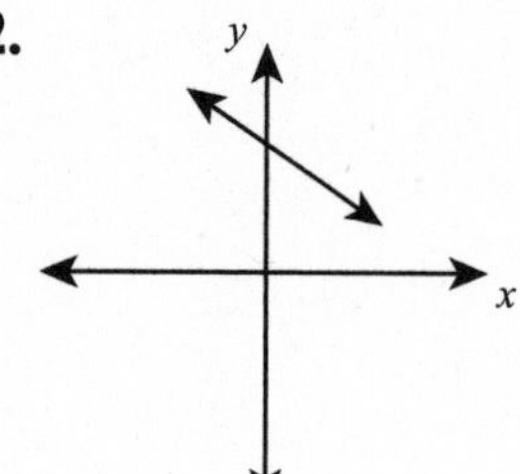

3.

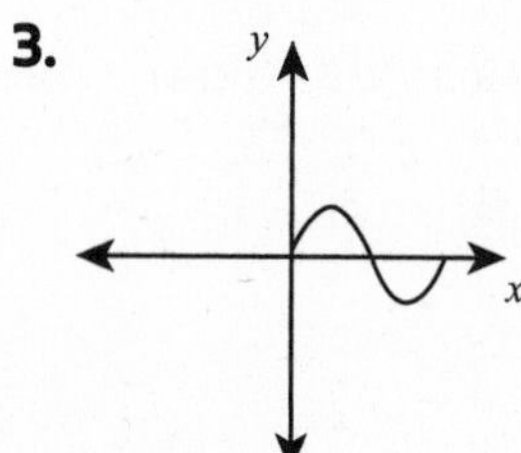

4.

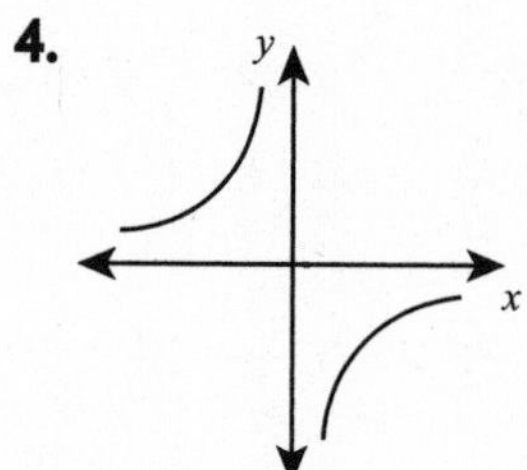

5.

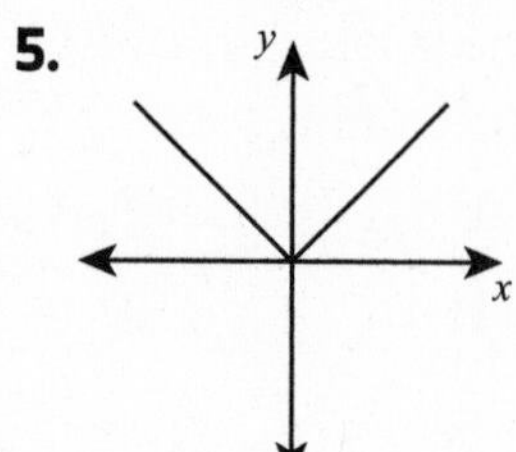

6.

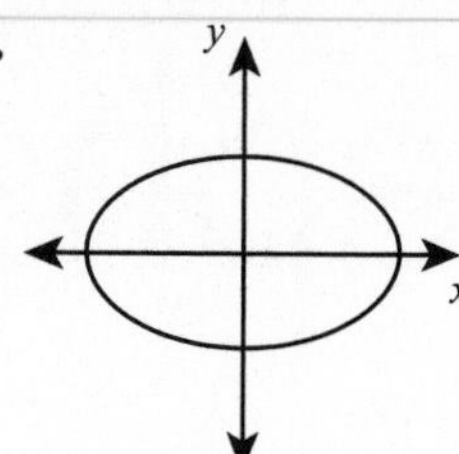

7.

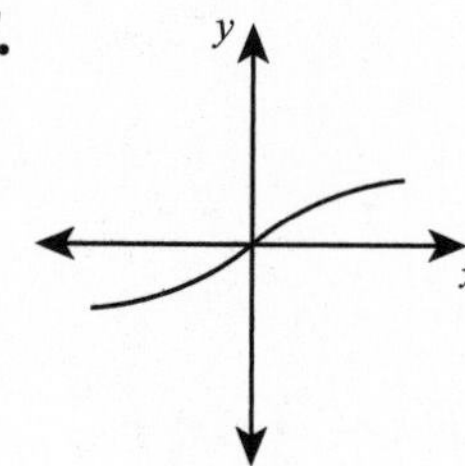

8.

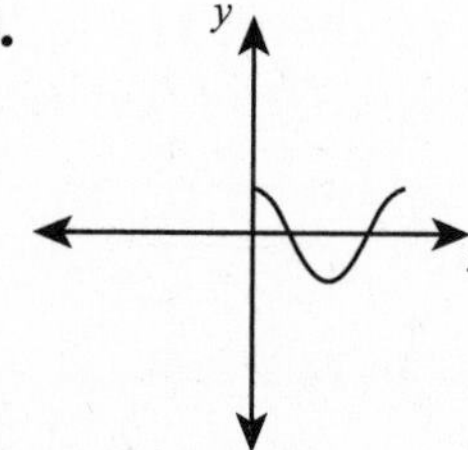

9.

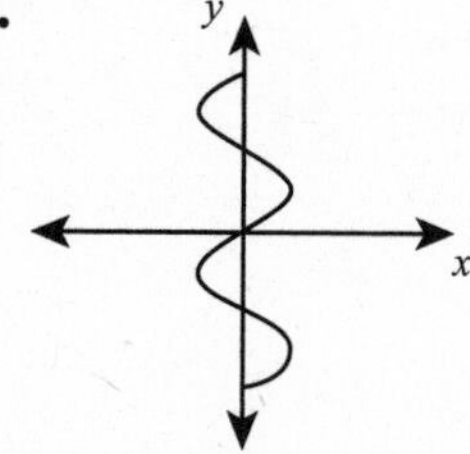

10.

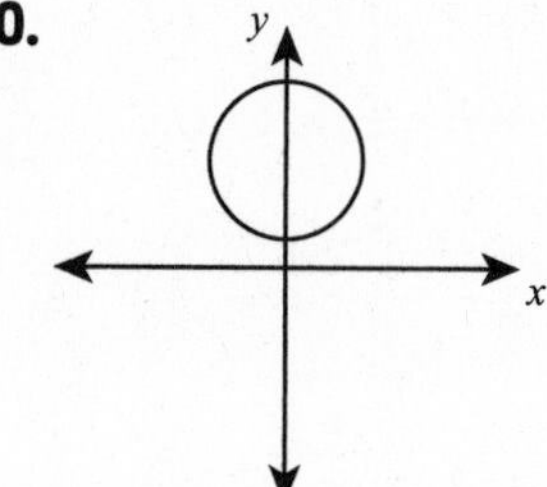

Zeros and Intercepts

A **zero** of a function f is an input value that produces a zero output value. Determine zeros of f by solving $f(x) = 0$. For functions whose outputs are real numbers, the zeros of the function are the same as the x-intercepts of the function's graph. An ***x*-intercept** of the graph of a function is the x-coordinate of a point at which the graph intersects the x-axis. A graph can have many x-intercepts, or it might not have any.

Only *real* zeros (if any) of a function are x-intercepts of its graph. Some functions have zeros that are not real numbers, so these zeros do not correspond to x-intercepts because these nonreal number values do not lie on the x-axis.

EXAMPLE

▶ For the function $f(x) = 2x - 3$, find its zeros and x-intercepts (if any).

Solve $2x - 3 = 0$.

$$2x - 3 = 0$$
$$2x - 3 + 3 = 0 + 3$$
$$2x = 3$$
$$\frac{2x}{2} = \frac{3}{2}$$
$$x = \frac{3}{2}$$

zero: $x = \frac{3}{2}$; x-intercept: $\frac{3}{2}$.

EXAMPLE

▶ For the function $g(x) = x^2 - x - 12$, find its zeros and x-intercepts (if any).

Solve $x^2 - x - 12 = 0$.

$$x^2 - x - 12 = 0$$
$$(x - 4)(x + 3) = 0$$
$$x - 4 = 0 \text{ or } x + 3 = 0$$
$$x = 4 \text{ or } x = -3$$

zeros: $x = 4$ and $x = -3$; x-intercepts: 4, -3.

EXAMPLE

For the function $h(x) = 2^x$, find its zeros and x-intercepts (if any).

Solve $2^x = 0$.

$2^x = 0$ has no solution. This is true because $2^x > 0$.

zeros: none; x-intercepts: none

A function f cannot have more than one y-intercept because, by definition, each x value in the domain of f is paired with *exactly one* y value in the range.

The ***y*-intercept** is the y-coordinate of the point at which the graph intersects the y-axis. The graph of a function has at most *one* y-intercept. If 0 is in the domain of f, then $f(0)$ is the y-intercept of the graph of f.

EXAMPLE

For the function $f(x) = 2x - 3$, find the y-intercept.

Determine $f(0)$.

$f(0) = 2(0) - 3 = -3$

y-intercept: -3

EXAMPLE

For the function $g(x) = x^2 - x - 12$, find the y-intercept.

Determine $g(0)$.

$g(0) = (0)^2 - 2(0) - 12 = -12$

y-intercept: -12

EXAMPLE

For the function $h(x) = 2^x$, find the y-intercept.

Determine $h(0)$.

$h(0) = 2^0 = 1$

y-intercept: 1

EXERCISE 30.2

For each function, (a) find the zeros and x-intercepts (if any), and (b) find the y-intercept.

1. $f(x) = 2x + 10$
2. $g(x) = x^2 + 5x + 6$
3. $h(x) = 1{,}500(2^x)$
4. $f(x) = x$
5. $g(x) = 5x^2$
6. $h(x) = 3^{x+1}$
7. $f(x) = \frac{4}{5}x$
8. $g(x) = 2x^2 - 5x - 3$
9. $h(x) = 5^{2x-3}$
10. $f(x) = -\frac{1}{4}x + 8$

Increasing, Decreasing, and Constant Behavior

A function is **increasing** on an interval if the output values increase as the input values increase over the interval. Similarly, a function is **decreasing** on an interval if the output values decrease as the input values increase over that interval. A function is **constant** on an interval if the output values stay constant as the input values increase over the interval.

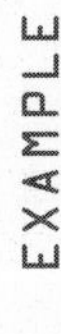

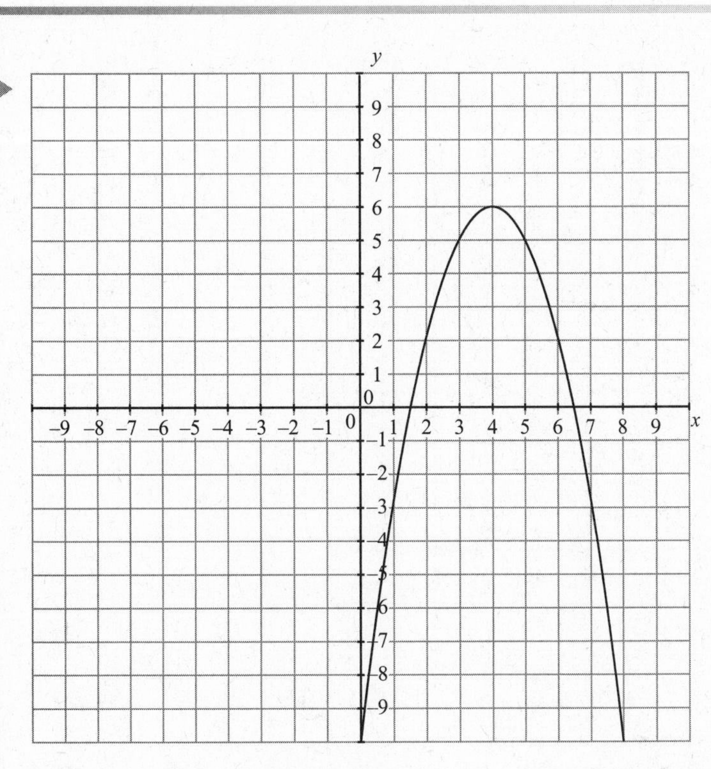

This function is increasing when $x < 4$ and decreasing when $x > 4$.

EXAMPLE

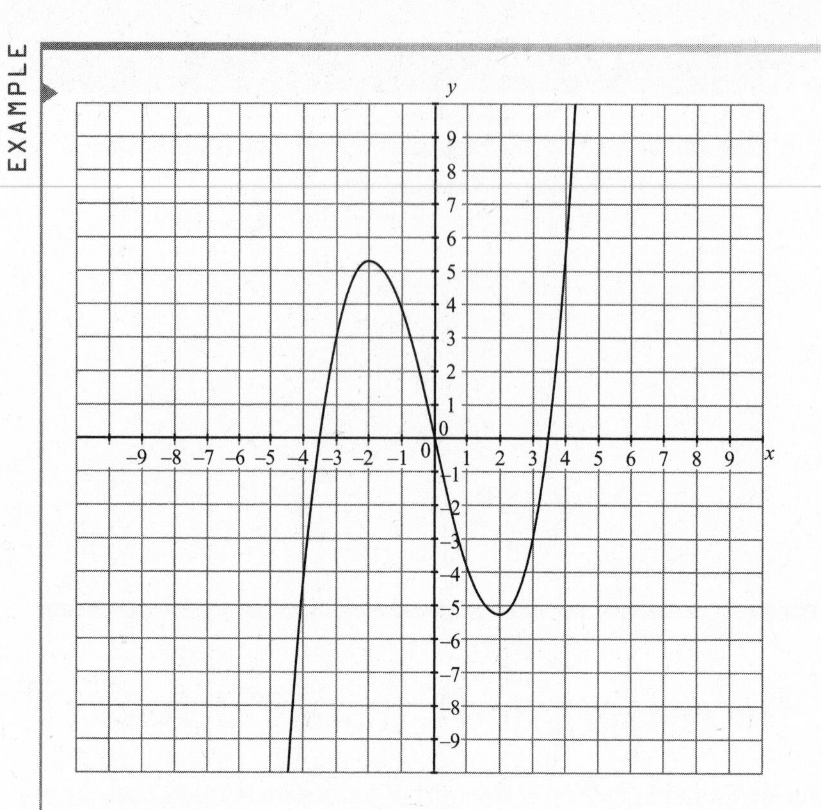

This function is increasing when $x < -2$, decreasing when x is between -2 and 2, and increasing when $x > 2$.

EXAMPLE

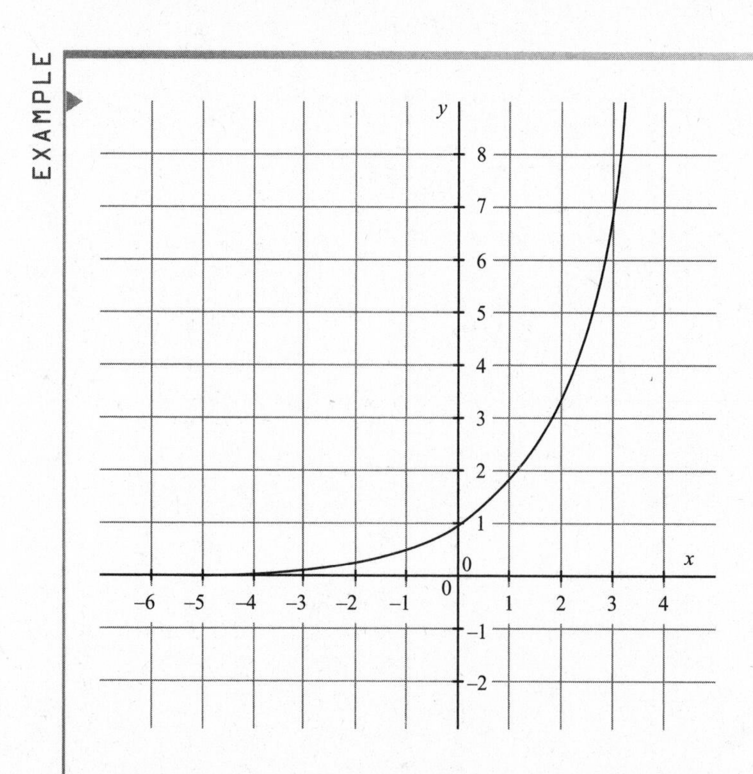

This function is increasing over its entire domain.

EXAMPLE

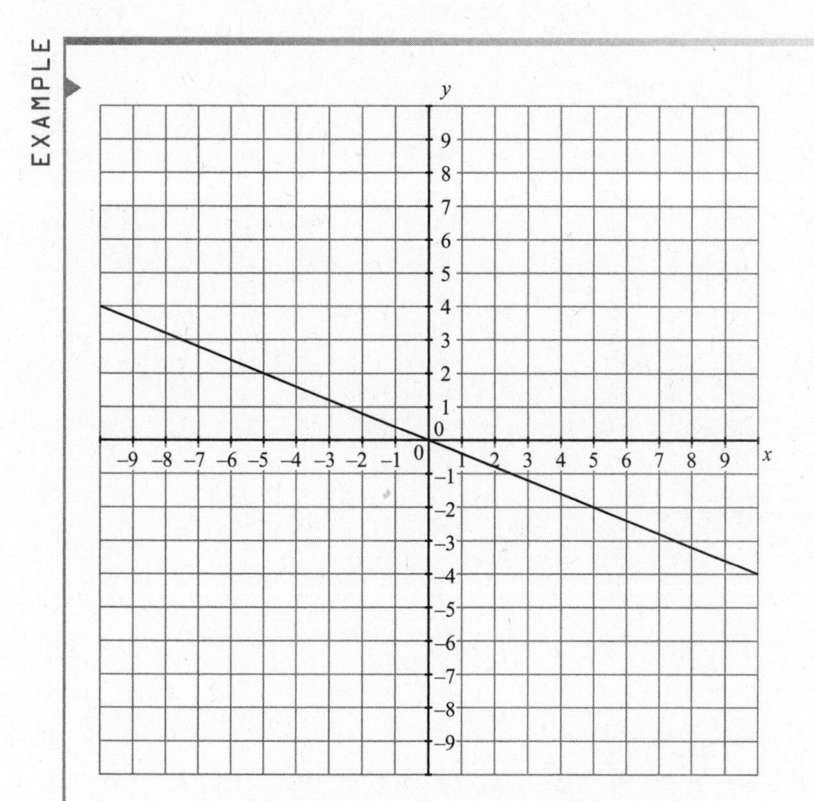

This function is decreasing over its entire domain.

An **increasing function** is one that is increasing over its entire domain, a **decreasing function** is one that is decreasing over its entire domain, and a **constant function** is one that is constant over its entire domain. Linear functions are either increasing, decreasing, or constant functions. Their graphs do not change direction.

EXAMPLE

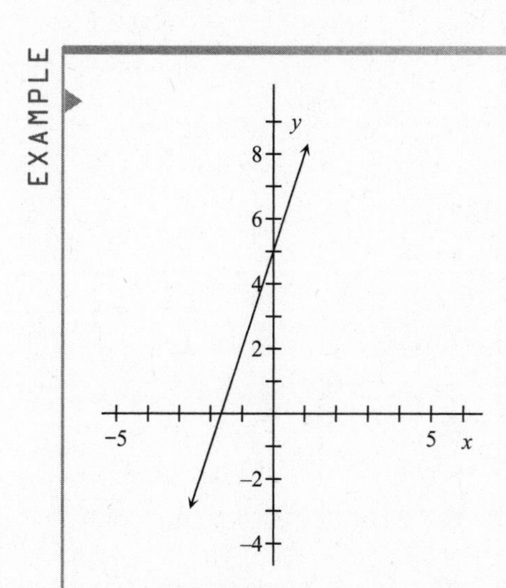

Increasing Linear Function

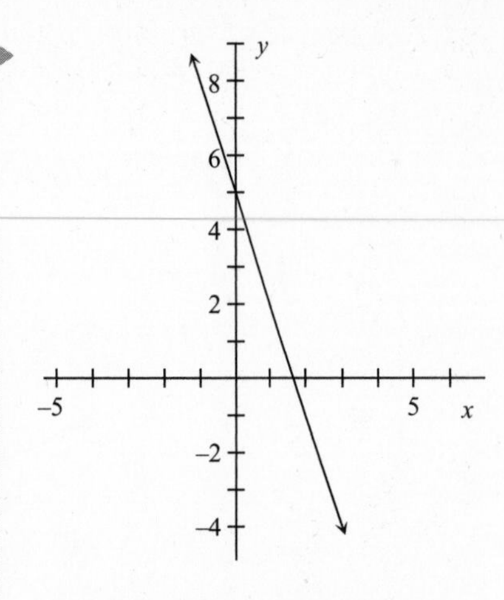

Decreasing Linear Function

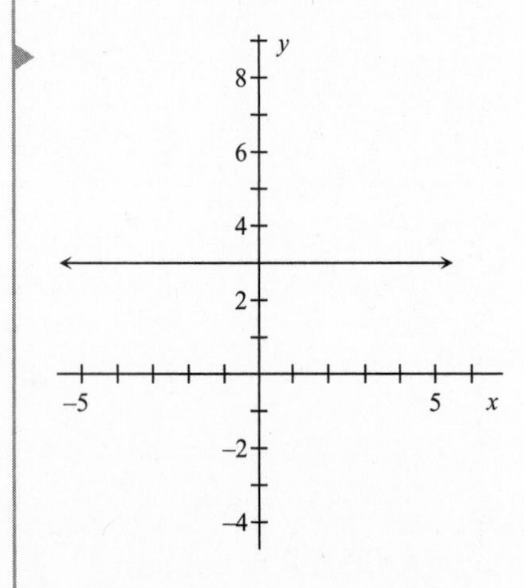

Constant Linear Function

EXERCISE 30.3

Describe the function's increasing, decreasing, and/or constant behavior.

1.

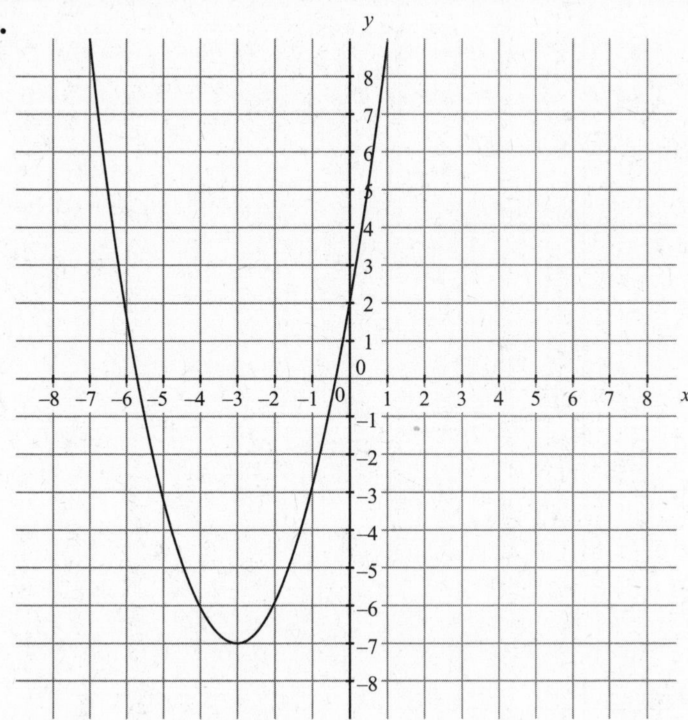

2.

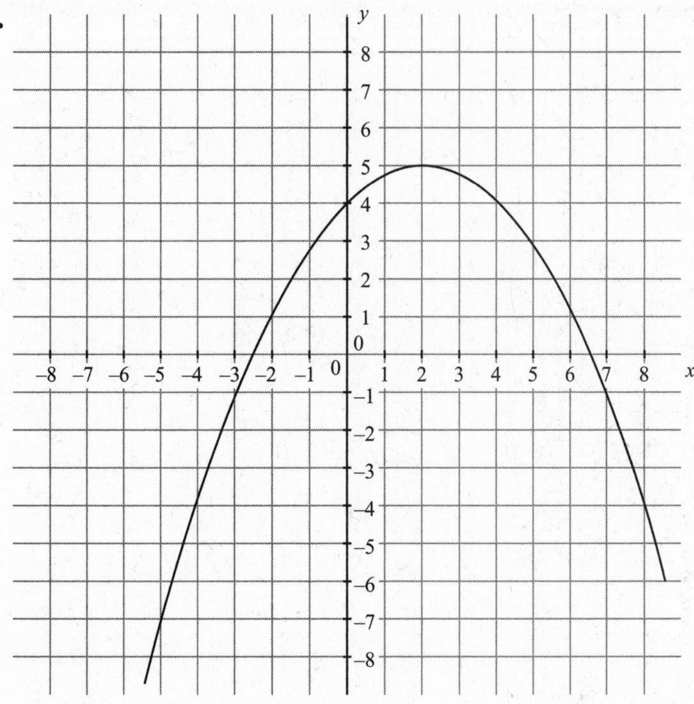

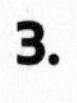

3.

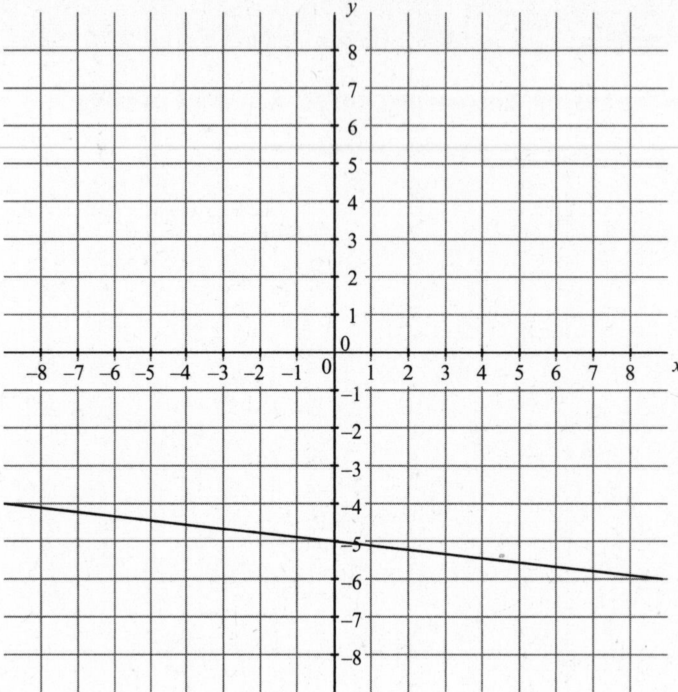
y
x
8
7
6
5
4
3
2
1
0
–8 –7 –6 –5 –4 –3 –2 –1 0 1 2 3 4 5 6 7 8
–1
–2
–3
–4
–5
–6
–7
–8

4.

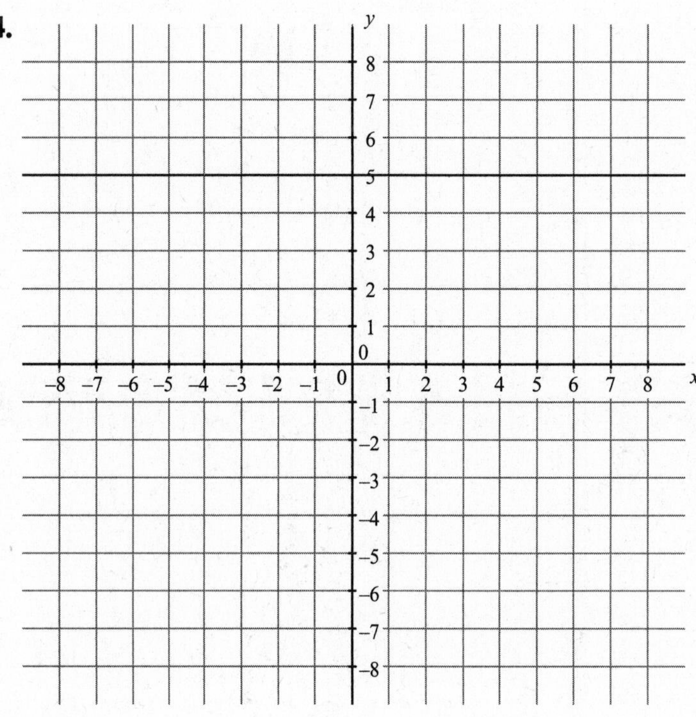
y
x
8
7
6
5
4
3
2
1
0
–8 –7 –6 –5 –4 –3 –2 –1 0 1 2 3 4 5 6 7 8
–1
–2
–3
–4
–5
–6
–7
–8

5.

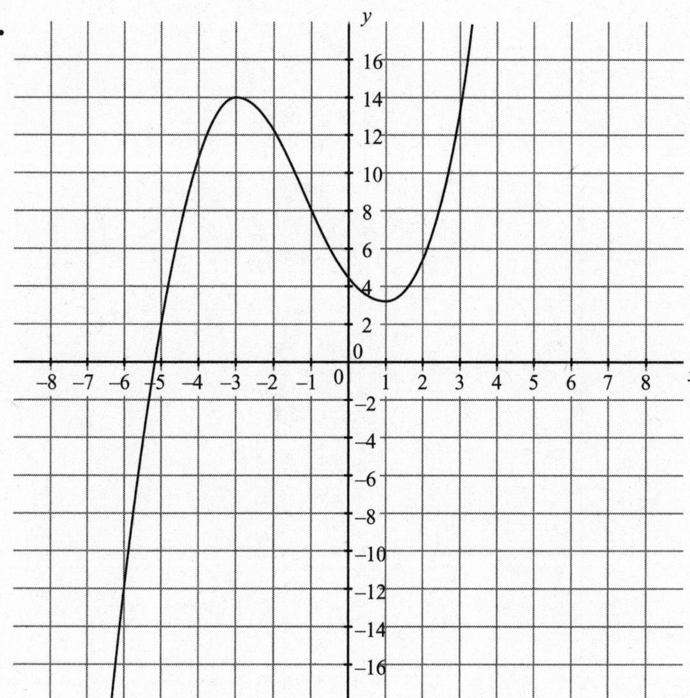
y
16
14
12
10
8
6
4
2
0
−2
−4
−6
−8
−10
−12
−14
−16
−8 −7 −6 −5 −4 −3 −2 −1 0 1 2 3 4 5 6 7 8
x

6.

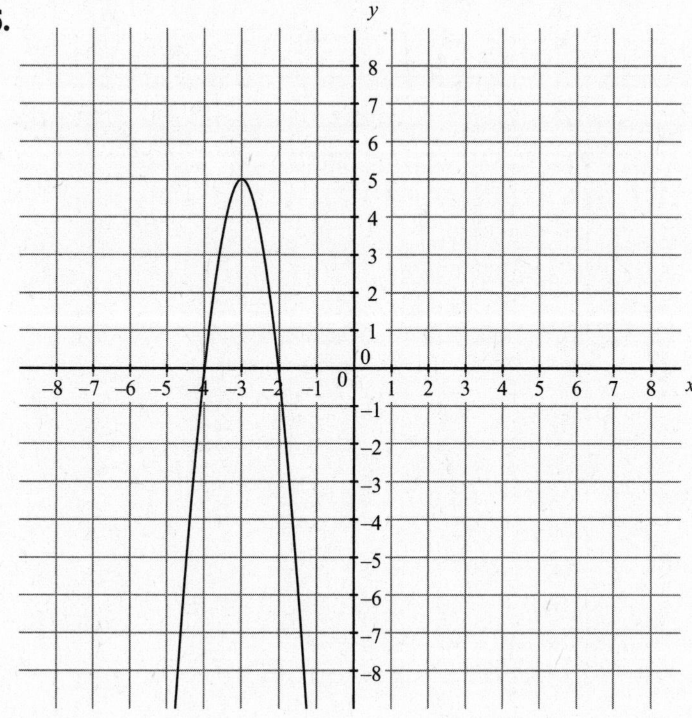
y
8
7
6
5
4
3
2
1
0
−1
−2
−3
−4
−5
−6
−7
−8
−8 −7 −6 −5 −4 −3 −2 −1 0 1 2 3 4 5 6 7 8
x

7.

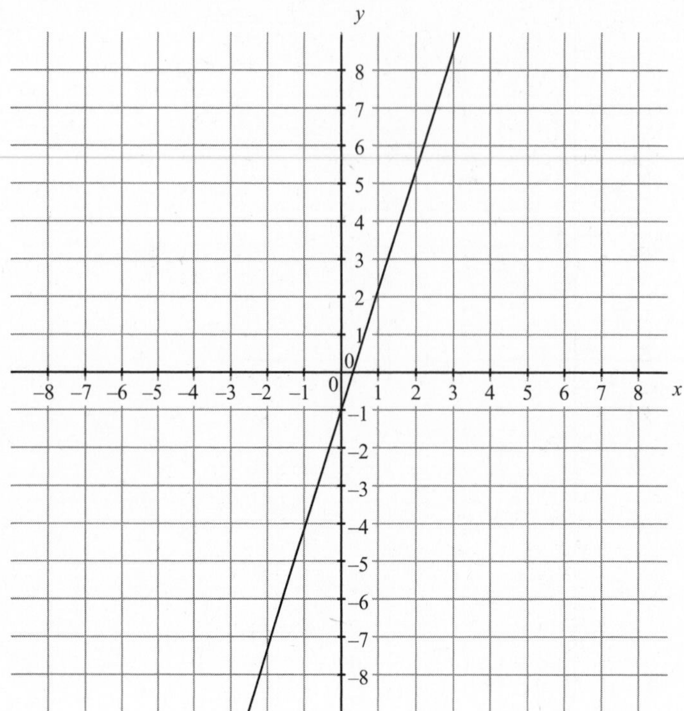
y
8
7
6
5
4
3
2
1
0
−8 −7 −6 −5 −4 −3 −2 −1 0 1 2 3 4 5 6 7 8 x
−1
−2
−3
−4
−5
−6
−7
−8

8.

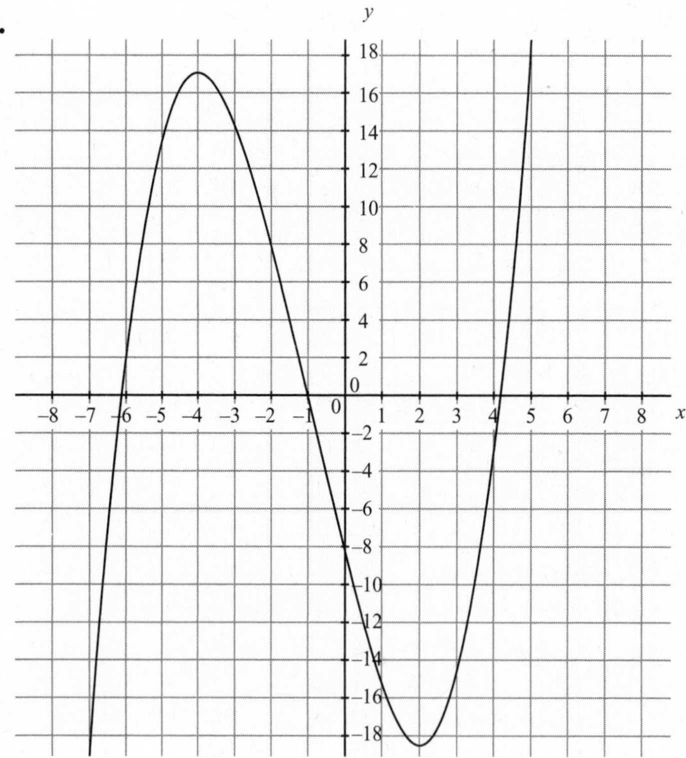
y
18
16
14
12
10
8
6
4
2
0
−8 −7 −6 −5 −4 −3 −2 −1 0 1 2 3 4 5 6 7 8 x
−2
−4
−6
−8
−10
−12
−14
−16
−18

9.

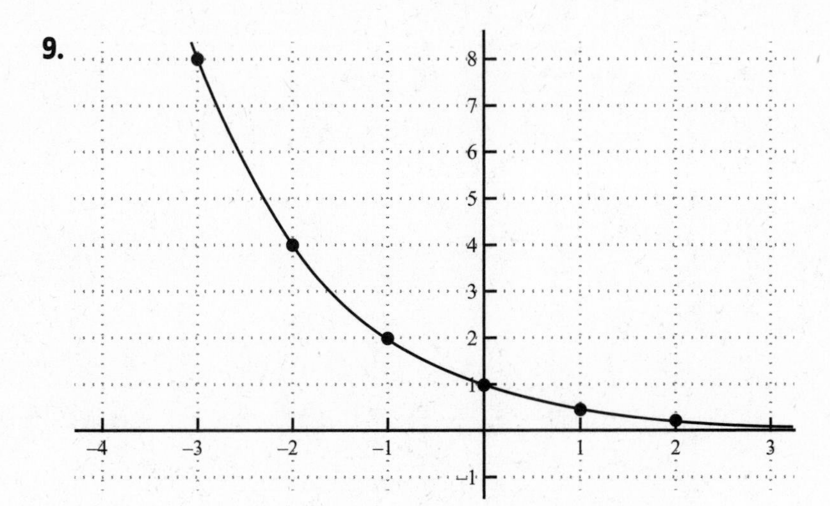

10.

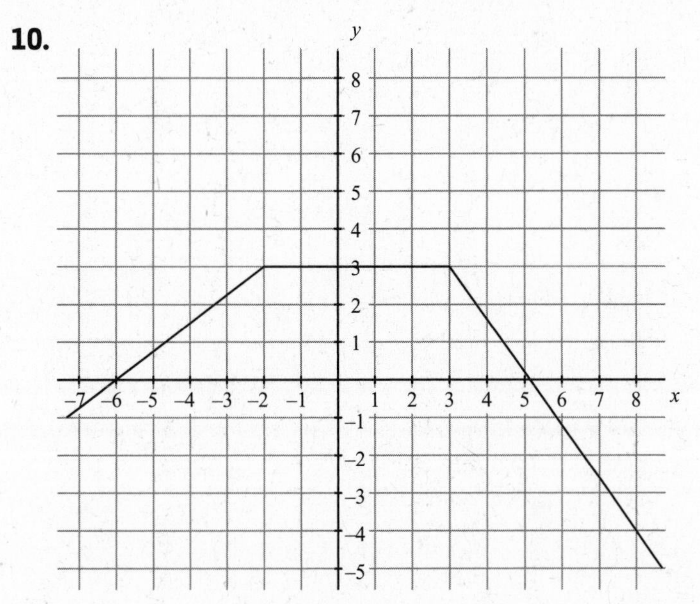

CHAPTER 31

Common Functions and Their Graphs

Linear Functions

The graph of the **linear function** $f(x) = mx + b$ is a nonvertical line with slope m and y-intercept b. When $m \neq 0$, the graph has exactly one y-intercept, b, and exactly one x-intercept, $-\frac{b}{m}$. Thus, the points $(0, b)$ and $\left(-\frac{b}{m}, 0\right)$ are contained in the graph. The only zero is the real number $-\frac{b}{m}$; thus, the graph crosses the x-axis at the point $\left(-\frac{b}{m}, 0\right)$. If $m > 0$, f is an increasing function; and if $m < 0$, f is a decreasing function.

When $m = 0$, $f(x) = b$ is the **constant function** and its graph is a horizontal line with slope 0 and y-intercept b. Constant functions either have no zeros or infinitely many zeros according to the following guideline: If $b \neq 0$, they have no zeros; if $b = 0$, every real number x is a zero.

EXAMPLE

The graph of $f(x) = -\frac{1}{2}x + 6$ is a nonvertical line with slope $-\frac{1}{2}$ and y-intercept 6. Its x-intercept is 12. The function f is a decreasing function.

EXAMPLE

The graph of $f(x) = x$ is a nonvertical line with slope 1 and y-intercept 0. Its x-intercept is 0. The function f is an increasing function.

The function $f(x) = x$ is the **identity function** (because it matches each input value with an identical output value).

EXAMPLE

The graph of $f(x) = 3x$ is a nonvertical line with slope 3 and y-intercept 0. Its x-intercept is 0. The function f is an increasing function.

Functions of the form $f(x) = kx$ are **directly proportional functions** and k is the nonzero **constant of proportionality**.

The slope m of a linear function is its **rate of change**. Because the slope of a line is constant, a linear function's rate of change is constant over its entire graph. The rate of change describes how the output changes in relation to the input. For every 1-unit change in the input, there are m units of change in the output. If the input changes by k units, the output changes by km units.

EXAMPLE

Given $f(x) = 2x + 5$, what is the rate of change of $f(x)$ with respect to x?

The rate of change equals 2, the slope of f.

EXAMPLE

Given $f(x) = 4x$, if the input changes 3 units, how many units does the output change?

The change in the output is $4(3) = 12$.

EXAMPLE

Given that the function $c(x) = 524 + 75x$ models the total cost of a hot water heater replacement, where $c(x)$ represents the total cost, in dollars, and x represents the number of hours to complete the replacement. What is the rate of change of the total cost with respect to the number of hours needed to complete the replacement?

The rate of change equals \$75 per hour.

EXERCISE 31.1

For 1 to 5, fill in the blanks to make true statements.

1. The graph of $f(x) = -4.5x + 9$ is a ____________ line with slope ____________ and y-intercept ____________. Its x-intercept is ____________. The function f is a(n) ____________ (increasing, decreasing) function.

2. The graph of $f(x) = 75x$ is a ____________ line with slope ____________ and y-intercept ____________. Its x-intercept is ____________. The function f is a(n) ____________ (increasing, decreasing) function.

3. The graph of $f(x) = -\frac{1}{5}x$ is a ____________ line with slope ____________ and y-intercept ____________. Its x-intercept is ____________. The function f is a(n) ____________ (increasing, decreasing) function.

4. The graph of $f(x) = 72 - 3x$ is a ____________ line with slope ____________ and y-intercept ____________. Its x-intercept is ____________. The function f is a(n) ____________ (increasing, decreasing) function.

5. The graph of $f(x) = -\frac{4}{3}x + 6$ is a ____________ line with slope ____________ and y-intercept ____________. Its x-intercept is ____________. The function f is a(n) ____________ (increasing, decreasing) function.

For 6-10, answer as indicated.

6. Given $f(x) = -3x + 4$, what is the rate of change of $f(x)$ with respect to x?

7. Given the function $f(x) = \frac{3}{5}x$, if the input changes 10 units, how many units does the output change?

8. The function $p(x) = 15x - 2{,}500$ models the net profit on sales of a school logo T-shirt, where $p(x)$ represents the net profit (in dollars) and x represents the number of T-shirts sold. What is the rate of change of the net profit with respect to the number of T-shirts sold?

9. The function $d(t) = 65t$ models the total distance (in miles) that a vehicle travels, where $d(t)$ represents the total distance traveled and t represents the number of hours traveled. What is the rate of change of the total distance traveled with respect to the number of hours traveled?

10. The function $h(t) = 1{,}000 - 8t$ models the number of gallons of water remaining in a 1,000-gallon tank after water has been draining for t hours. What is the rate of change of the height of the water with respect to the number of hours elapsed?

An **absolute minimum** of a function's graph is the least output value obtained over the entire domain of the function.
An **absolute maximum** of a function's graph is the greatest output value obtained over the entire domain of the function.

Quadratic Functions

The graph of the **quadratic function** $f(x) = ax^2 + bx + c$ $(a \neq 0)$ is a parabola. The vertex is $\left(-\frac{b}{2a}, f\left(-\frac{b}{2a}\right)\right)$. The y-intercept is $f(0) = c$. When $a > 0$, the parabola opens upward and the y-coordinate of the vertex is the absolute minimum of f. When $a < 0$, the parabola opens downward and the y-coordinate of the vertex is the absolute maximum of f. The parabola is symmetric about its axis of symmetry, a vertical line, with the equation $x = -\frac{b}{2a}$, through its vertex that is parallel to the y-axis.

EXAMPLE

The graph of $f(x) = x^2 - x - 6$ is a parabola with vertex $\left(\frac{1}{2}, -\frac{25}{4}\right)$. The y-intercept is -6. The graph's curve opens upward and the y-coordinate of the vertex is an absolute minimum of f. The graph is symmetric about the equation $x = \frac{1}{2}$.

EXAMPLE

▶ The graph of $f(x) = -x^2 + 6x - 9$ is a parabola with vertex (3, 0). The y-intercept is -9. The graph's curve opens downward and the y-coordinate of the vertex is an absolute maximum of f. The graph is symmetric about the equation $x = 3$.

EXAMPLE

▶ The graph of $f(x) = x^2 - 9$ is a parabola with vertex (0, -9). The y-intercept is -9. The graph's curve opens upward and the y-coordinate of the vertex is an absolute minimum of f. The graph is symmetric about the equation $x = 0$.

The zeros of $f(x) = ax^2 + bx + c$ are the roots of the quadratic equation $ax^2 + bx + c = 0$. Depending on the solution set of $ax^2 + bx + c = 0$, the graph of a quadratic function might or might not have x-intercepts. Three cases occur:

- If there are two real *unequal* roots, the parabola will have x-intercepts at those two values.
- If there is exactly one real root, the parabola will have an x-intercept at only that one value.
- If there are no real roots, the parabola will *not* have any x-intercepts.

EXAMPLE

▶ Find the real zeros and x-intercepts (if any) of $f(x) = x^2 - x - 6$.

Solve $x^2 - x - 6 = 0$.

$$
\begin{aligned}
x^2 - x - 6 &= 0 \\
(x + 2)(x - 3) &= 0 \\
x + 2 = 0 \text{ or } x - 3 &= 0 \\
x = -2 \text{ or } x &= 3
\end{aligned}
$$

zeros: $x = -2$ or $x = 3$; x -intercepts: -2, 3.

EXAMPLE

Find the real zeros and x-intercepts (if any) of $f(x) = -x^2 + 6x - 9$.

Solve $-x^2 + 6x - 9 = 0$.

$$\begin{aligned} -x^2 + 6x - 9 &= 0 \\ x^2 - 6x + 9 &= 0 \\ (x-3)^2 &= 0 \\ x - 3 &= 0 \\ x &= 3 \end{aligned}$$

zero: $x = 3$; x -intercept: 3.

EXAMPLE

Find the real zeros and x-intercepts (if any) of $f(x) = x^2 + x + 1$.

Solve $x^2 + x + 1 = 0$.

$a = 1$, $b = 1$, and $c = 1$

$$x = \frac{-b \pm \sqrt{b^2 - 4ac}}{2a} = \frac{-(1) \pm \sqrt{(1)^2 - 4(1)(1)}}{2(1)} = \frac{1 \pm \sqrt{1-4}}{2} = \frac{1 \pm \sqrt{-3}}{2}$$
$$= \text{no real number roots}$$

real zeros: none; x -intercepts: none

To graph $f(x) = ax^2 + bx + c$, plot its vertex, its x- and y-intercepts. Sketch a smooth U-shaped curve through the points. If $a > 0$, draw the curve opening upward. If $a < 0$, draw the curve opening downward.

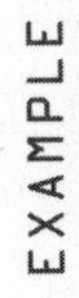

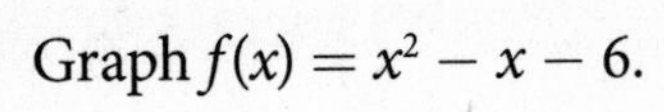

Graph $f(x) = x^2 - x - 6$.

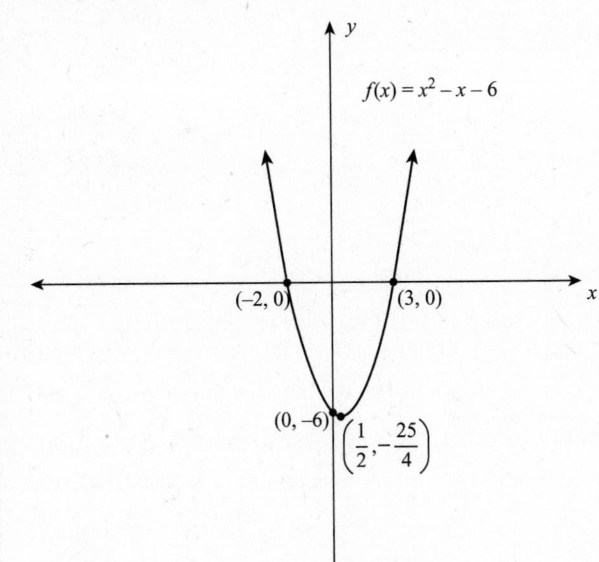

EXAMPLE

Graph $f(x) = -x^2 + 6x - 9$.

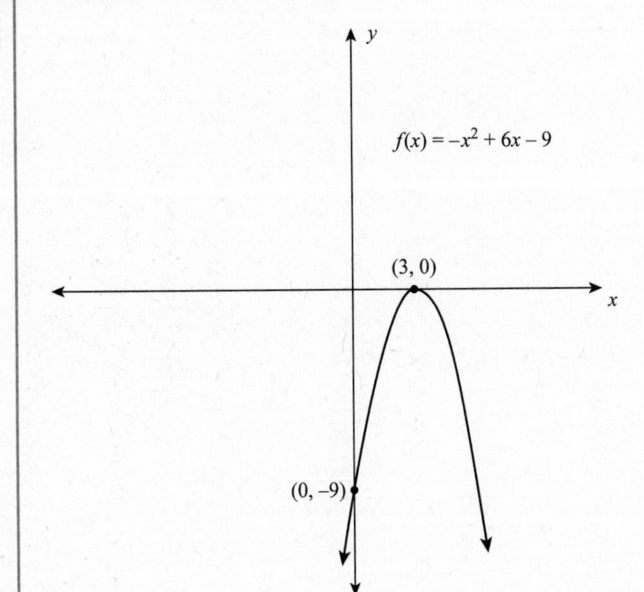

EXERCISE 31.2

For 1 to 4, fill in the blanks to make true statements.

1. The graph of $f(x) = 2x^2 - 5x - 3$ is a __________ with vertex __________. The y-intercept is __________. The graph's curve opens __________ and the y-coordinate of the vertex is an absolute __________ of f. The graph is symmetric about the equation __________.

2. The graph of $f(t) = -t^2 + 100$ is a __________ with vertex __________. The y-intercept is __________. The graph's curve opens __________ and the y-coordinate of the vertex is an absolute __________ of f. The graph is symmetric about the equation __________.

3. The graph of $f(x) = x^2 + 10x + 25$ is a __________ with vertex __________. The y-intercept is __________. The graph's curve opens __________ and the y-coordinate of the vertex is an absolute __________ of f. The graph is symmetric about the equation __________.

4. The graph of $f(x) = -x^2 - 2x + 8$ is a __________ with vertex __________. The y-intercept is __________. The graph's curve opens __________ and the y-coordinate of the vertex is an absolute __________ of f. The graph is symmetric about the equation __________.

For 5 to 9, find the real zeros and x-intercepts (if any), of the quadratic function.

5. $f(x) = 2x^2 - 5x - 3$
6. $f(t) = -t^2 + 100$
7. $f(x) = x^2 + 10x + 25$
8. $f(x) = -x^2 - 2x + 8$
9. $f(x) = x^2 - 9$

For 10 to 11, graph the function.

10. $f(x) = 2x^2 - 5x - 3$
11. $f(x) = -x^2 - 2x + 8$

In 12 to 15, given that the function $h(t) = -16t^2 + 64t + 36$ models the height t in feet of a ball after t seconds of elapsed time, answer the following questions.

12. What is the height of the ball after 3 seconds?
13. At what elapsed time will the ball reach maximum height?
14. What is the maximum height attained by the ball?
15. At what time after the ball is in the air will its height be zero?

Exponential Functions

The function $f(x) = b^x$ ($b \neq 1, b > 0$) is the **exponential function**, with base b. Because $b^x > 0$ for every real number x, the exponential function's graph does not cross the x-axis. Thus, there are no zeros and no x-intercepts. The y-intercept is $f(0) = b^0 = 1$. The graph passes through the points (0, 1) and (1, b) and is located in the first and second quadrants only.

EXAMPLE

The following figure shows the graph of the exponential function $f(x) = 2^x$.

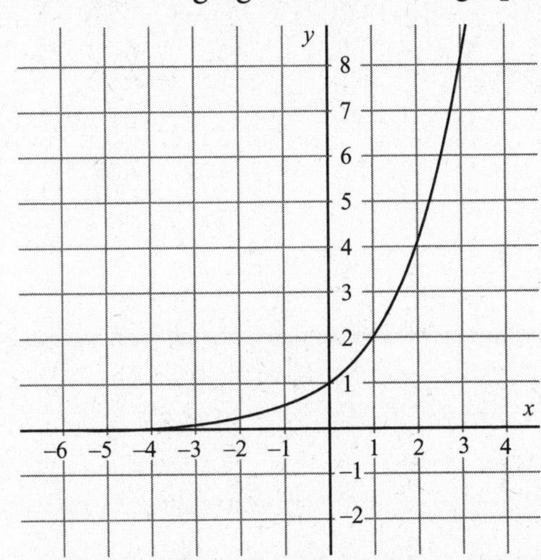

Two important exponential functions are $f(x) = 10^x$ with base 10; and $f(x) = e^x$, the **natural exponential function**, with base e.

EXERCISE 31.3

For 1 to 3, evaluate the function using $f(x) = 5^x$.

1. $f(-2)$

2. $f(0)$

3. $f(2)$

For 4 to 6, evaluate the function using $f(x) = 4^{-x}$.

4. $f(-2)$

5. $f(0)$

6. $f(2)$

For 7 to 9, evaluate the function using $f(x) = \left(\frac{1}{2}\right)^x$.

7. $f(-2)$

8. $f(0)$

9. $f(2)$

For 10 to 12, evaluate the function using $f(t) = 1{,}000\,(2)^{t/4}$.

10. $f(0)$

11. $f(4)$

12. $f(20)$

For 13 to 15, answer as indicated.

13. The function $b(t) = 50(2^t)$ models the growth of a bacteria population (in millions) over t hours of elapsed time. What is the number of bacteria after 3 hours?

14. The function $g(t) = 1{,}500(1.02)^t$ models the growth g of an investment (in dollars) over a period of t years. What is the value of the investment after 2 years?

15. The function $v(t) = 25{,}000(.75)^t$ models the depreciated value v (in dollars) of a used vehicle over a period of t years. To the nearest cent, what is the value of the vehicle after 5 years?

CHAPTER 32

Introduction to Systems of Two-Variable Equations

Definition and Terminology for Systems of Two-Variable Linear Equations

A **system of two linear equations in two variables** consists of a pair of linear equations in the same two variables. To **solve a system** of linear equations in two variables means to find all ordered pairs of values for the two variables that make *both* equations true simultaneously. An ordered pair that makes an equation true **satisfies** the equation. When an ordered pair makes both equations in a system true, the ordered pair **satisfies** the system.

EXAMPLE

Determine whether the given ordered pair satisfies the system.

$$\begin{matrix} y = 2x - 5 \\ x = 1 - y \end{matrix}, (2,-1)$$

Check whether (2,–1) makes both equations true.

$$y = 2x - 5$$
$$\rightarrow (-1) \stackrel{?}{=} 2(2) - 5$$
$$-1 = -1 \text{ True}$$

$$x = 1 - y$$
$$(2) \stackrel{?}{=} 1 - (-1)$$
$$2 = 2 \text{ True}$$

The ordered pair (2,–1) satisfies the system because it makes both equations true simultaneously.

EXAMPLE

Determine whether the given ordered pair satisfies the system.

$$\begin{array}{l} 3x - y = 1 \\ x + y = -5 \end{array}, (1,2)$$

Check whether (1,2) makes both equations true.

$3x - y = 1$ | $x + y = -5$

$3(1) - (2) \stackrel{?}{=} 1$ | $1 + 2 \stackrel{?}{=} -5$

$1 = 1$ True | $3 = -5$ False

The ordered pair (1,2) does not satisfy the system because it does not satisfy both equations.

When you are solving a system of two linear equations in two variables, the standard form of writing them together is as follows:

$$\begin{array}{l} A_1x + B_1y = C_1 \\ A_2x + B_2y = C_2 \end{array}$$

EXAMPLE

Write the system in standard form.

$$\begin{array}{l} y = 2x - 5 \\ x = 1 - y \end{array} \rightarrow \begin{array}{l} y - 2x = -5 \\ x + y = 1 \end{array}$$

$$\begin{array}{l} y = 3x - 1 \\ 6x - 2y = -2 \end{array} \rightarrow \begin{array}{l} -3x + y = -1 \\ 6x - 2y = -2 \end{array}$$

$$\begin{array}{l} y - 6 = 2x \\ \frac{1}{2}y = x + 3 \end{array} \rightarrow \begin{array}{l} y - 2x = 6 \\ -x + \frac{1}{2}y = 3 \end{array}$$

EXERCISE 32.1

For 1 to 5, state Yes or No as to whether the given ordered pair satisfies the system. Justify your answer.

1. $\begin{array}{l} x - 2y = -4 \\ 2x + y = 7 \end{array}, (-2,1)$

2. $\begin{array}{l} 4x - y = 3 \\ x - 3y = -13 \end{array}, (2,5)$

3. $\begin{array}{l} 4x + 2y = 8 \\ 2x - 3y = -8 \end{array}, \left(\frac{1}{2},3\right)$

4. $\begin{array}{l} -x + 2y = 4 \\ 2x + y = 7 \end{array}, (0,2)$

5. $\begin{array}{l} 8x - 2y = 24 \\ 4x - y = 12 \end{array}, (3,0)$

For 6 to 10, write the system in standard form.

6. $-2y = 6 - 8x$
   $x = 3y - 13$

7. $y = 4 - 2x$
   $2x = 3y - 8$

8. $4x = 8 - 2y$
   $2x + y = -8$

9. $3 - 2y = -3x$
   $6x + 2y - 9 = 0$

10. $14y + 7x = 2$
    $14x - 7y = -11$

Types of Solutions for Systems of Two-Variable Linear Equations

The **solution set** of a system is the collection of all solutions. There are three possibilities: the system has *exactly one solution, no solution,* or *infinitely many solutions.*

A system of two linear equations in standard form has exactly one solution if $\frac{A_1}{A_2} \neq \frac{B_1}{B_2}$; infinitely many solutions if $\frac{A_1}{A_2} = \frac{B_1}{B_2} = \frac{C_1}{C_2}$; or no solution if $\frac{A_1}{A_2} = \frac{B_1}{B_2} \neq \frac{C_1}{C_2}$. *Before* you attempt to solve a system of equations, check the system's coefficient ratios to determine its type of solution set.

In many cases, you can mentally check a system's coefficient ratios.

EXAMPLE

Determine the system's type of solution set.

$-2x + y = -5$
$x + y = 1$
$\rightarrow$ Exactly one solution because $\frac{-2}{1} \neq \frac{1}{1}$.

$-2x + y = -5$
$4x - 2y = 10$
$\rightarrow$ Infinitely many solutions because $\frac{-2}{4} = \frac{1}{-2} = \frac{-5}{10}$.

$-2x + y = -5$
$10x - 5y = 15$
$\rightarrow$ No solution because $\frac{-2}{10} = \frac{1}{-5} \neq \frac{-5}{15}$

Geometrically, you can represent the two equations of a system of linear equations in two variables as two lines in the coordinate plane. For the two lines, three possibilities can occur, corresponding to the three possibilities for the solution set. If the system has exactly one solution, then the two lines *intersect* in a unique point in the plane. The ordered pair that corresponds to the point of intersection is the solution to the system. If the system has infinitely many solutions, then the two lines are *coincident* (that is, have all points in common). If the system has no solutions, then the two lines are *parallel* in the plane.

EXAMPLE

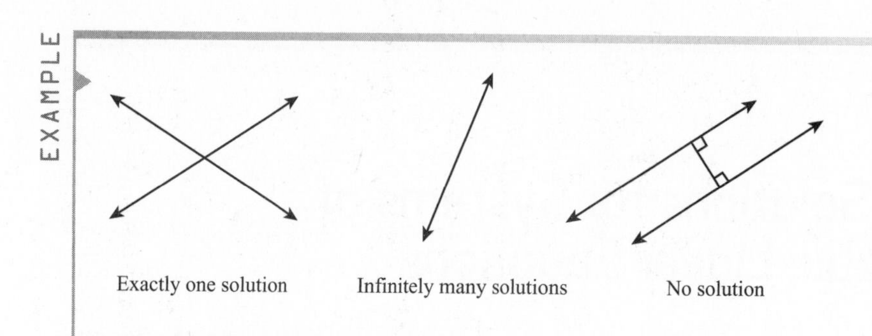

EXERCISE 32.2

For 1 to 5, state whether the system has exactly one solution, no solution, or infinitely many solutions.

1. $2x + y = 4$
$2x - 3y = -8$

2. $4x + 2y = 8$
$2x + y = -8$

3. $3x - 2y = -3$
$6x + 2y = 9$

4. $7x + 14y = 2$
$14x - 7y = -11$

5. $2x - y = 4$
$4x - 2y = 8$

For 6 to 10, state whether the two lines representing the given system are intersecting, coincident, or parallel.

6. $2x + y = 4$
$2x - 3y = -8$

7. $4x + 2y = 8$
$2x + y = -8$

8. $3x - 2y = -3$
$6x + 2y = 9$

9. $7x + 14y = 2$
$14x - 7y = -11$

10. $2x - y = 4$
$4x - 2y = 8$

CHAPTER 33

Solving Systems of Two-Variable Linear Equations

Solving Systems of Two-Variable Linear Equations by Substitution

To solve a system of linear equations in standard form by substitution, do the following:

1. Solve one equation for one of the variables in terms of the other variable.
2. In the other equation, replace the variable solved for in Step 1 with the expression obtained, simplify, and solve for the second variable.
3. Using the first equation, substitute the value obtained in Step 2 for the second variable, simplify, and solve for the first variable.
4. State the solution set.

You should check your solution set in the system's original equations.

Use your judgment to decide which variable to solve for first.

EXAMPLE

Solve the system by substitution.

$$2x - y = 0$$
$$x + y = 3$$

Solve the first equation, $2x - y = 0$, for y in terms of x.

$$2x - y = 0$$
$$2x - y + y = 0 + y$$
$$2x = y$$

Substitute $2x$ for y into the second equation, $x + y = 3$, and solve for x.

$$x + (2x) = 3$$
$$x + 2x = 3$$
$$3x = 3$$
$$\frac{3x}{3} = \frac{3}{3}$$
$$x = 1$$

When solving by substitution, enclose substituted values in parentheses to avoid errors.

Substitute 1 for x into the second equation, $x + y = 3$, and solve for y. (You can substitute the value for x into either equation.)

$$x + y = 3$$
$$1 + y = 3$$
$$1 + y - 1 = 3 - 1$$
$$y = 2$$

The solution set is the ordered pair (1, 2).

EXERCISE 33.1

Solve the system. For 1 to 14, Solve the system by substitution.

1. $x - 2y = -4$
$2x + y = 7$

2. $4x - y = 3$
$x - 3y = -13$

3. $4x + 2y = 8$
$2x - 3y = -8$

4. $-x + 2y = 4$
$2x + y = 7$

5. $8x - 2y = 24$
$4x - y = 12$

6. $8x - 2y = 6$
$x - 3y = -13$

7. $2x + y = 4$
$2x - 3y = -8$

8. $4x + 2y = 8$
$2x + y = -8$

9. $3x - 2y = -3$
$6x + 2y = 9$

10. $7x + 14y = 2$
$14x - 7y = -11$

11. $2x - y = 4$
$4x - 2y = 8$

12. $5x + 2y = 3$
$2x + 3y = -1$

13. $7x - 14y = 2$
$x - 2y = 1$

14. $2x = y + 4$
$x + y = 5$

For 15, answer as indicated.

15. The sum of two numbers is 25 and their difference is 2. What are the two numbers?

Solving Systems of Two-Variable Linear Equations by Elimination

To solve a system of linear equations in standard form by elimination, do the following:

1. Choosing either variable as a target for elimination, if necessary, multiply one or both of the equations by a nonzero constant or constants to make that variable's coefficients sum to zero.
2. Add the transformed equations to eliminate the target variable.
3. Solve for the variable that was not eliminated.
4. Substitute the value obtained in Step 3 into one of the original equations, simplify, and solve for the other variable.
5. State the solution set.

You can choose either variable as the target variable. Use your judgment to decide.

EXAMPLE

Solve the system by elimination.

$$2x - y = 4$$
$$x + 2y = -3$$

To eliminate x, multiply the second equation by -2.

$$2x - y = 4$$
$$x + 2y = -3$$

$$\begin{array}{lcr} 2x - y = 4 & \xrightarrow{\hspace{2cm}} & 2x - y = 4 \\ x + 2y = -3 & \xrightarrow[\text{Multiply by } -2]{} & -2x - 4y = 6 \end{array}$$

Add the resulting two equations.

$$\begin{array}{r} 2x - y = 4 \\ -2x - 4y = 6 \\ \hline -5y + 0 = 10 \end{array}$$

Solve $-5y = 10$ for y.

$$-5y = 10$$
$$\frac{-5y}{-5} = \frac{10}{-5}$$
$$y = -2$$

Substitute -2 for y into one of the original equations, $2x - y = 4$, and solve for x.

$$2x - y = 4$$
$$2x - (-2) = 4$$
$$2x + 2 = 4$$
$$2x + 2 - 2 = 4 - 2$$
$$2x = 2$$
$$\frac{2x}{2} = \frac{2}{2}$$
$$x = 1$$

The solution set is the ordered pair $(1, -2)$.

EXERCISE 33.2

For 1 to 14, Solve the system by elimination.

1. $x - 2y = -4$
$2x + y = 7$

2. $4x - y = 3$
$x - 3y = -13$

3. $4x + 2y = 8$
$2x - 3y = -8$

4. $-x + 2y = 4$
$2x + y = 7$

5. $8x - 2y = 24$
$4x - y = 12$

6. $8x - 2y = 6$
$x - 3y = -13$

7. $2x + y = 4$
$2x - 3y = -8$

8. $4x + 2y = 8$
$2x + y = -8$

9. $3x - 2y = -3$
$6x + 2y = 9$

10. $7x + 14y = 2$
$14x - 7y = -11$

11. $2x - y = 4$
$4x - 2y = 8$

12. $5x + 2y = 3$
$2x + 3y = -1$

13. $7x - 14y = 2$
$x - 2y = 1$

14. $2x = y + 4$
$x + y = 5$

For 15, answer as indicated.

15. In a group of sheep and chickens, there are 84 eyes and 122 legs. How many animals of each type are there?

Solving Systems of Two-Variable Linear Equations by Graphing

To solve a system of linear equations by graphing do the following:

1. Graph the two equations.
2. Locate (as accurately as possible) the intersection point on the graph.
3. Check whether the ordered pair from Step 2 satisfies both equations.
4. State the solution set.

The graphing method might yield inaccurate results due to limitations of graphing.

EXAMPLE

Solve the system by graphing.

$3x - y = 1$
$x + y = -5$

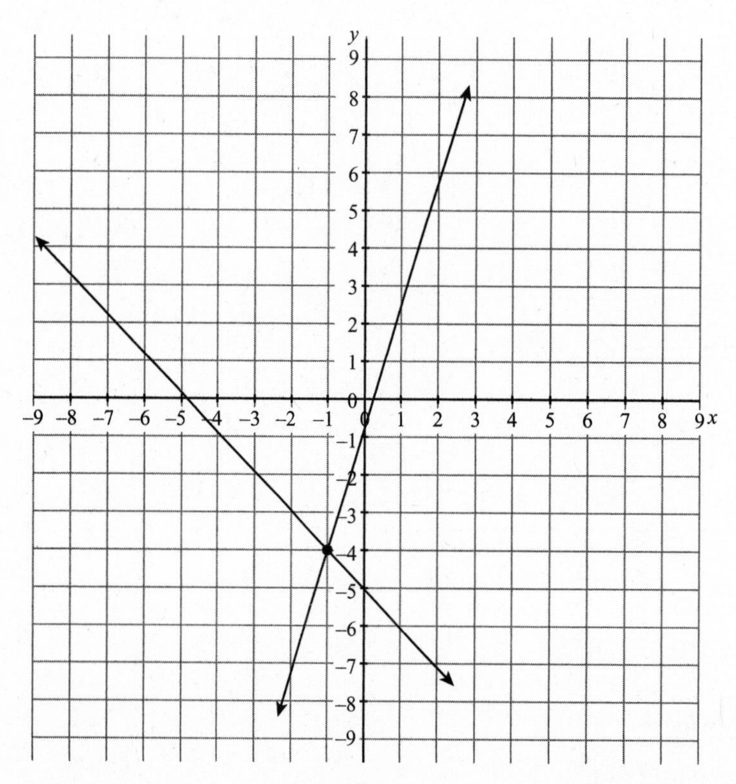

The lines appear to intersect at (−1, −4). Check (−1, −4) in both equations.

$3x - y = 1$

$3(-1) - (-4) \stackrel{?}{=} 1$

$-3 + 4 \stackrel{?}{=} 1$

$1 = 1$ True

$x + y = -5$

$(-1) + (-4) \stackrel{?}{=} -5$

$-5 = -5$ True

The solution set is (−1, −4) because it satisfies both equations.

EXERCISE 33.3

Solve the system by graphing.

1. $x - 2y = -4$
$2x + y = 7$

2. $4x - y = 3$
$x - 3y = -13$

3. $4x + 2y = 8$
$2x - 3y = -8$

4. $-x + 2y = 4$
$2x + y = 7$

5. $8x - 2y = 6$
$x - 3y = -13$

6. $2x + y = 4$
$2x - 3y = -8$

7. $4x + 2y = 5$
$2x - y = 1$

8. $5x + 2y = 3$
$2x + 3y = -1$

9. $3x - 2y = -3$
$6x + 2y = 9$

10. $3x - 2y = 5$
$6x - 4y = 11$

CHAPTER 34

Graphing Systems of Two-Variable Inequalities

Graphing Two-Variable Linear Inequalities

The graph of a two-variable linear inequality is the half-plane that contains all solutions to the inequality. To graph the inequality, do the following: Transform the inequality into an equivalent form in which you have only y on the left side of the inequality symbol. Next, replace the inequality symbol with an equal sign. Then graph the resulting linear function to obtain a boundary line for the inequality. Use a dashed line for $<$ and $>$, and a solid line for $\leq$ and $\geq$. If the inequality contains $<$ or $\leq$, shade the portion of the plane beneath the boundary line. If the inequality $>$ or $\geq$, shade the portion of the plane above the boundary line.

EXAMPLE

Graph $3x + y \geq 4$.

Rewrite $3x + y \geq 4$ as $y \geq -3x + 4$. Graph $y = -3x + 4$. Use a solid line and shade the portion of the plane above the line.

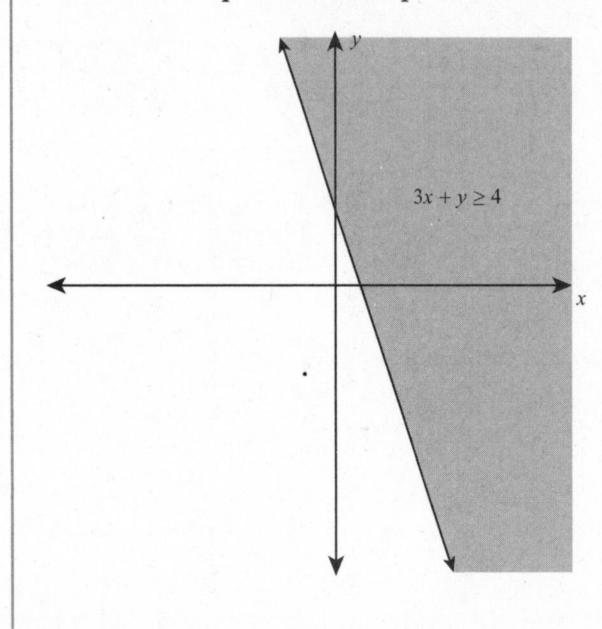

EXAMPLE

Graph $-5x + 2y < 6$.

Rewrite $-5x + 2y < 6$ as $y < \frac{5}{2}x + 3$. Graph $y = \frac{5}{2}x + 3$. Use a dashed line and shade the portion of the plane below the boundary line.

y
x
$-5x + 2y < 6$

EXERCISE 34.1

For 1 to 5, state whether the half-plane Above or Below the boundary line is shaded in the graph of the linear inequality.

1. $x + y > -5$

2. $\frac{3}{2}x - y > 5$

3. $x - y \geq 4$

4. $4x - 3y \leq 12$

5. $4 - y > 0$

For 6 to 10, graph the linear inequality.

6. $x + y > -5$

7. $\frac{3}{2}x - y > 5$

8. $x - y \geq 4$

9. $4x - 3y \leq 12$

10. $4 - y > 0$

Graphing Two-Variable Quadratic Inequalities

The graph of a two-variable quadratic inequality is the half-plane that contains all solutions to the inequality. To graph the inequality, do the following: Transform the inequality into an equivalent form in which you have only y on the left side of the inequality symbol. Next, replace the inequality symbol with an equal sign. Then graph the resulting quadratic function to obtain a boundary parabola for the inequality. Use a dashed curve for $<$ and $>$, and a solid curve for $\leq$ and $\geq$. If the inequality contains $<$ or $\leq$, shade the portion of the plane beneath the boundary parabola. If the inequality $>$ or $\geq$, shade the portion of the plane above the boundary parabola.

EXAMPLE

Graph $y + 4 > x^2 - x - 2$.

Rewrite $y + 4 > x^2 - x - 2$ as $y > x^2 - x - 6$. Graph $y = x^2 - x - 6$. Use a dashed line and shade the portion of the plane above the boundary parabola.

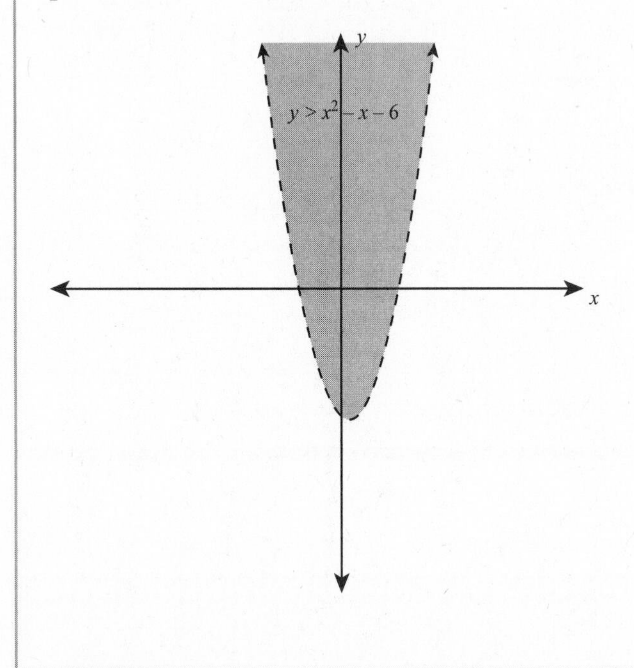

EXAMPLE

Graph $-x^2 + 6x - 9 \geq y$.

Rewrite $-x^2 + 6x - 9 \geq y$ as $y \leq -x^2 + 6x - 9$. Graph $y = -x^2 + 6x - 9$. Use a solid line and shade the portion of the plane below the boundary parabola.

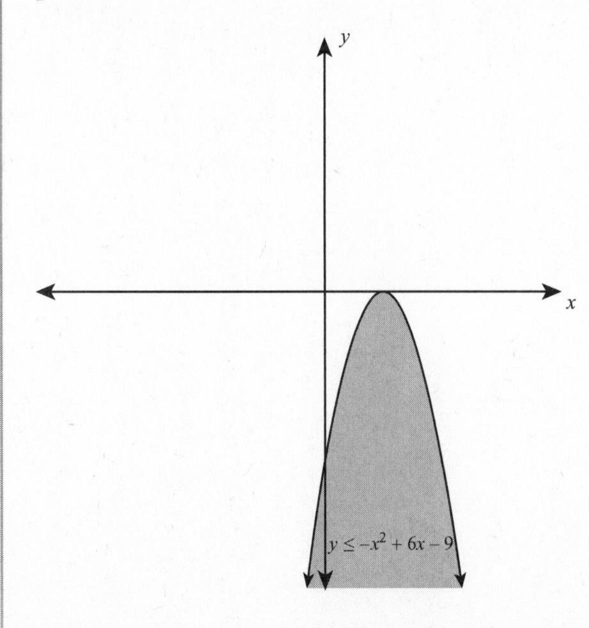

EXERCISE 34.2

For 1 to 5, state whether the half-plane Above or Below the boundary parabola is shaded in the graph of the quadratic inequality.

1. $y > 4x^2$

2. $2x^2 + y \leq 8$

3. $x^2 > y + 1$

4. $2x^2 - 5x < y + 3$

5. $x^2 - y \leq 9$

For 6 to 10, graph the quadratic inequality.

6. $y > 4x^2$

7. $2x^2 + y \leq 8$

8. $x^2 > y + 1$

9. $2x^2 - 5x < y + 3$

10. $x^2 - y \leq 9$

APPENDIX A

Measurement Units and Conversions

U.S. Customary Units	Conversion
Length	
Inch (in)	$1 \text{ in} = \frac{1}{12} \text{ ft}$
Foot (ft)	1 ft = 12 in
Yard (yd)	1 yd = 36 in 1 yd = 3 ft
Mile (mi)	1 mi = 5,280 ft 1 mi = 1,760 yd
Weight	
Pound (lb)	1 lb = 16 oz
Ton (T)	1 T = 2,000 lb
Capacity	
Fluid ounce (fl oz)	$1 \text{ fl oz} = \frac{1}{8} \text{ c}$
Cup (c)	1 c = 8 fl oz
Pint (pt)	1 pt = 2 c
Quart (qt)	1 qt = 32 fl oz 1 qt = 4 c 1 qt = 2 pt
Gallon (gal)	1 gal = 128 fl oz 1 gal = 16 c 1 gal = 8 pt 1 gal = 4 qt

Metric Units	Conversion
Length	
Millimeter (mm)	$1\ \text{mm} = 0.001\ \text{m} = \frac{1}{1000}\ \text{m}$
Centimeter (cm)	$1\ \text{cm} = 10\ \text{mm}$
Meter (m)	$1\ \text{m} = 1000\ \text{mm}$ $1\ \text{m} = 100\ \text{cm}$
Kilometer (km)	$1\ \text{km} = 1000\ \text{m}$
Mass	
Milligram (mg)	$1\ \text{mg} = 0.001\ \text{g} = \frac{1}{1000}\ \text{g}$
Gram (g)	$1\ \text{g} = 1000\ \text{mg}$
Kilogram (kg)	$1\ \text{kg} = 1000\ \text{g}$
Capacity	
Milliliter (mL)	$1\ \text{mL} = 0.001\ \text{L} = \frac{1}{1000}\ \text{L}$
Liter (L)	$1\ \text{L} = 1000\ \text{mL}$

Time	Conversion
Second (s)	$1\ \text{s} = \frac{1}{60}\ \text{min}$
Minute (min)	$1\ \text{min} = 60\ \text{s}$
Hour (hr)	$1\ \text{hr} = 3600\ \text{s}$ $1\ \text{hr} = 60\ \text{min}$
Day (d)	$1\ \text{d} = 24\ \text{hr}$
Week (wk)	$1\ \text{wk} = 7\ \text{d}$
Year (yr)	$1\ \text{yr} = 365\ \text{d}$ $1\ \text{yr} = 52\ \text{wk}$

APPENDIX B

Geometry Formulas

Triangle

height h
base b
sides a, b, and c

$\text{Area} = \frac{1}{2}bh$

$\text{Perimeter} = a + b + c$
Sum of the measures of the interior angles $= 180°$

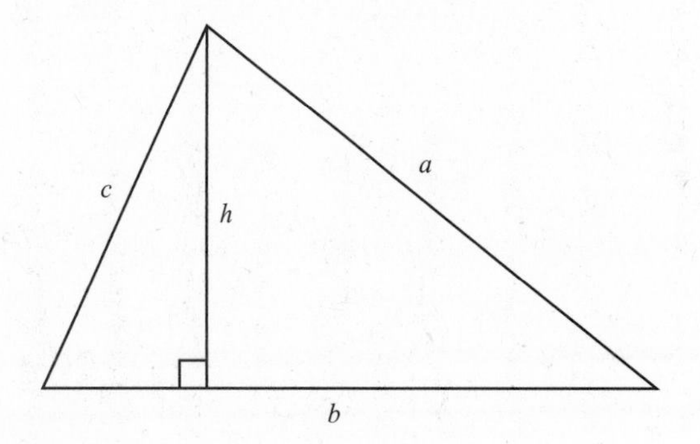

Right Triangle

Pythagorean theorem: $a^2 + b^2 = c^2$

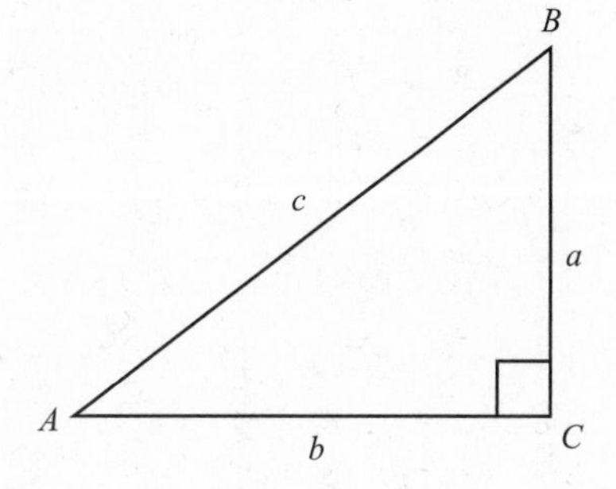

Square

side s
$\text{Area} = s^2$
$\text{Perimeter} = 4s$

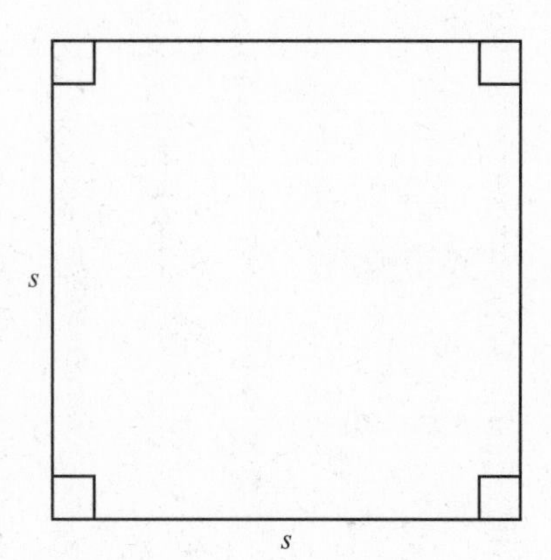

Rectangle

length l
width w
Area $= lw$
Perimeter $= 2l + 2w = 2(l + w)$

Parallelogram

height h
base b
width a
Area $= bh$
Perimeter $= 2a + 2b = 2(a + b)$

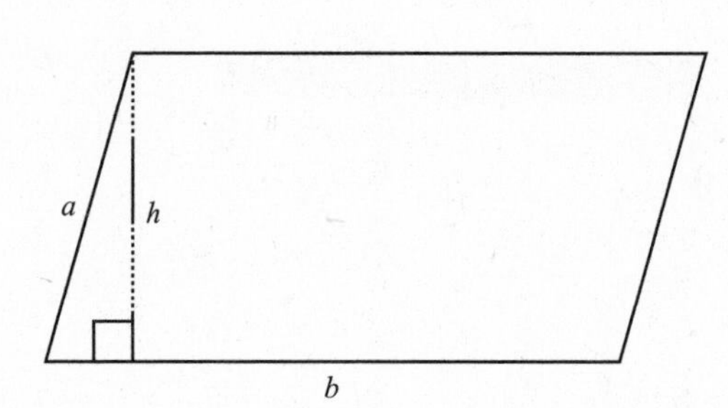

Circle

radius r
diameter d
Area $= \pi r^2$
Circumference $= 2\pi r = \pi d$
Diameter $d = 2r$

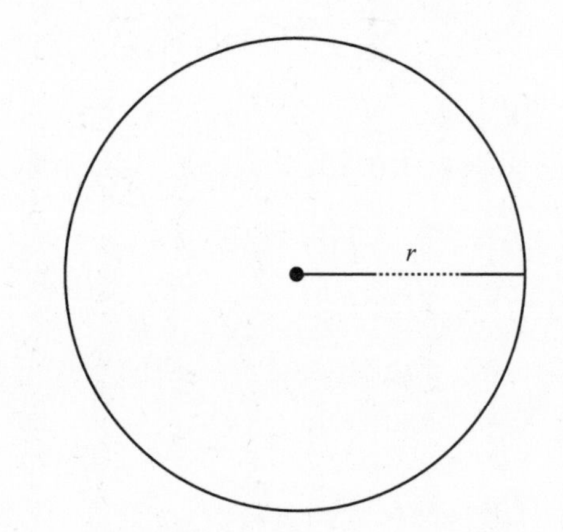

Trapezoid

height h
bases a, b

$$\text{Area} = \frac{1}{2}h(a+b)$$

Perimeter $= a + b + c + d$

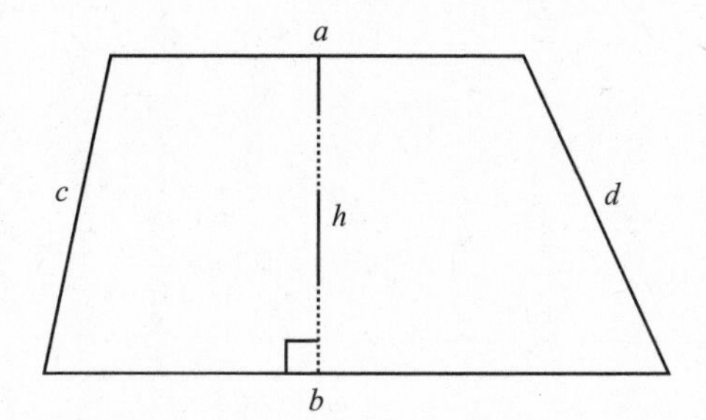

Sphere

radius r

$$\text{Volume} = \frac{4}{3}\pi r^3$$

Surface area $= 4\pi r^2$

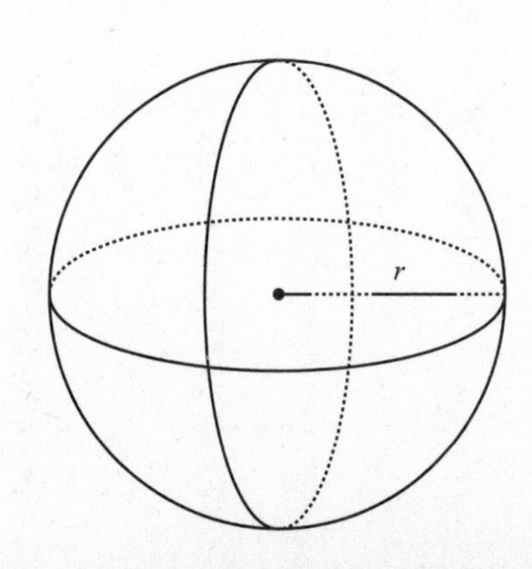

Right Prism

height h
area of base B
Volume $= Bh$
Total surface area $= 2B +$ sum of areas of rectangular sides

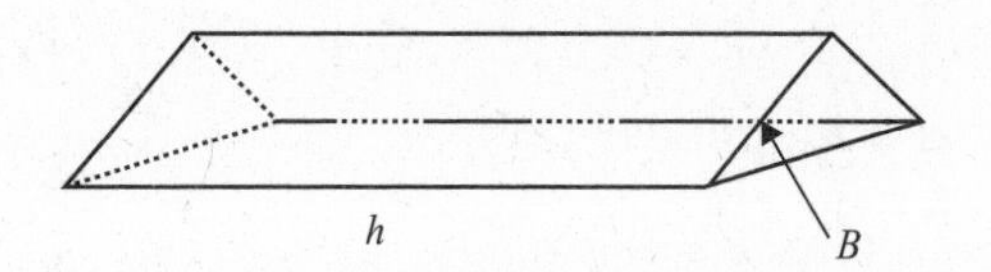

Right Rectangular Prism

length l
width w
height h
Volume $= lwh$
Total surface area $= 2hl + 2hw + 2lw$

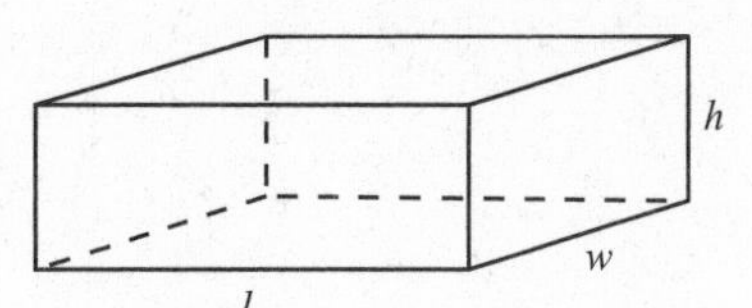

Cube

edge s
Volume $= s^3$
Total surface area $= 6s^2$

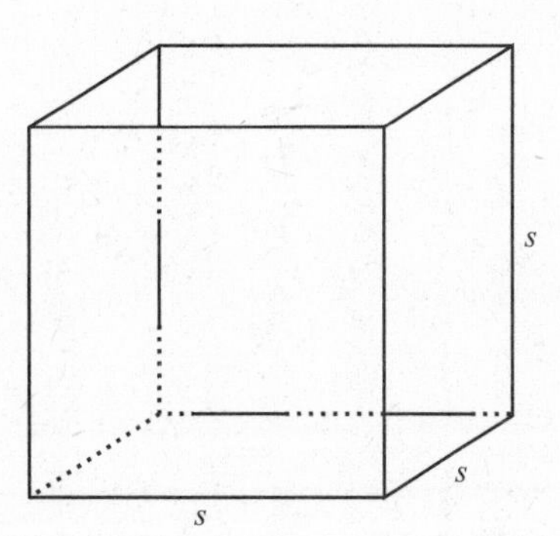

Right Circular Cylinder

height h
radius of base r
Volume $= \pi r^2 h$
Total surface area $= (2\pi r)h + 2(\pi r^2)$

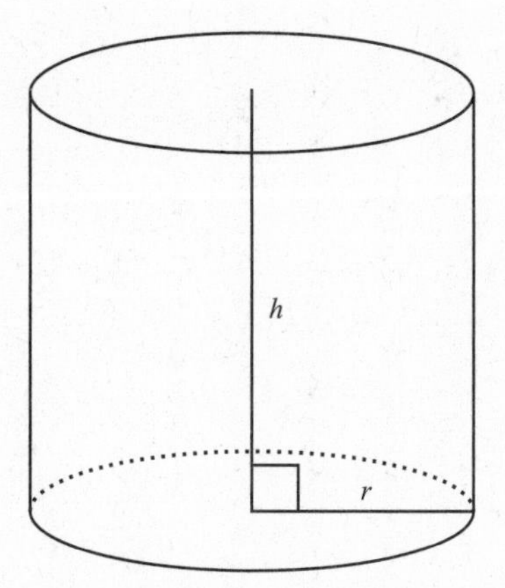

Right Pyramid

height h
area of base B
volume $= \frac{1}{3} Bh$
Total surface area $= B +$ sum of areas of triangular lateral faces

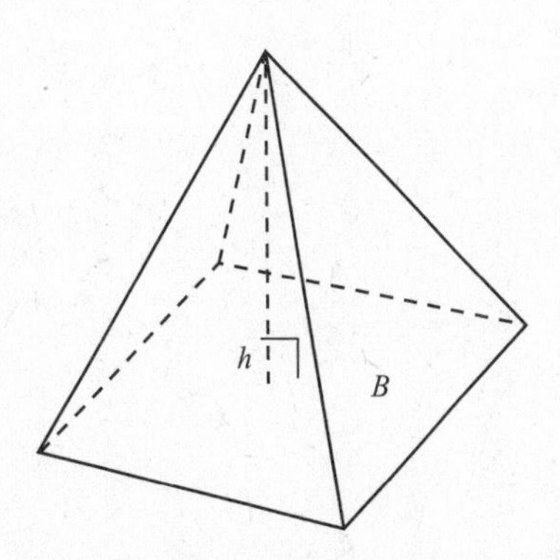

Right Circular Cone

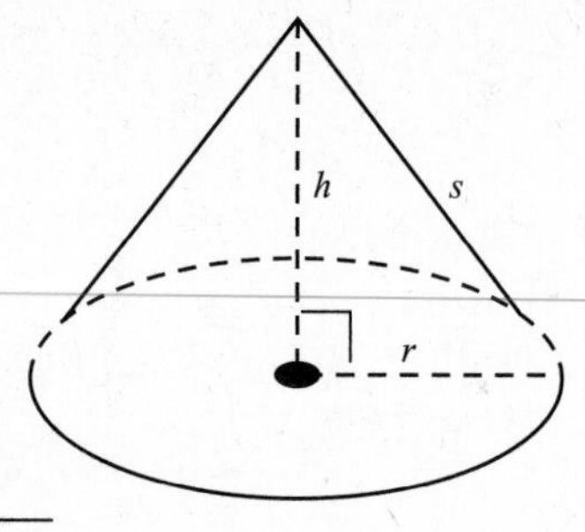

height h
radius of base r

Volume $= \dfrac{1}{3}\pi r^2 h$

Total surface area $= \pi r\sqrt{r^2 + h^2} + \pi r^2 = \pi rs + \pi r^2$,

where the slant height $s = \sqrt{r^2 + h^2}$

Answer Key

CHAPTER 1

Understanding the Real Numbers

EXERCISE 1.1

1. Rational
2. Rational
3. Rational
4. Rational
5. Rational
6. Irrational
7. Rational
8. Rational
9. Rational
10. Rational
11. Rational
12. Rational
13. Rational
14. Rational
15. Irrational
16. Rational
17. Rational
18. Rational
19. Rational
20. Rational

EXERCISE 1.2

1. Irrational, because 5 is not a perfect square.
2. Rational, because $\sqrt[3]{8} = 2$.
3. Rational, because $\sqrt{4} = 2$.
4. Rational, because $-\sqrt{4} = -2$.
5. Not real, because square roots of negative numbers are not real numbers.
6. Rational, because $\sqrt{16} = 4$.
7. Rational, because $\sqrt[3]{125} = 5$.

8. Rational, because $\sqrt[3]{-125} = -5$.
9. Irrational, because 12 is not a perfect fourth power.
10. Rational, because $\sqrt[5]{32} = 2$.
11. Rational, because $\sqrt[5]{-32} = -2$.
12. Rational, because $\sqrt{0.25} = 0.5$
13. Rational, because $\sqrt{\frac{9}{4}} = \frac{3}{2}$.
14. Rational, because $\sqrt[3]{64} = 4$.
15. Rational, because $\sqrt{100} = 10$.
16. Irrational, because $\frac{5}{3}$ is not a perfect square.
17. Irrational, because $-\frac{1}{3}$ is not a perfect cube.
18. Rational, because $-\sqrt{625} = -25$.
19. Not a real number, because even roots of negative numbers are not real numbers.
20. Irrational, because 75 is not a perfect cube.

CHAPTER 2

Properties of the Real Numbers

EXERCISE 2.1

1. Associative property of multiplication
2. Commutative property of addition
3. Commutative property of multiplication
4. Associative property of addition
5. Commutative property of multiplication
6. Commutative property of addition
7. Commutative property of addition
8. Associative property of addition
9. Commutative property of multiplication
10. Commutative property of multiplication
11. $2 \cdot \frac{3}{4} = \underline{\frac{3}{4} \cdot 2}$
12. $(7 \cdot 8)5 = \underline{7(8 \cdot 5)}$
13. $\sqrt{19} + \sqrt{3} = \underline{\sqrt{3} + \sqrt{19}}$
14. $4 + 2 + 16 + 8 = \underline{2 + 4} + 16 + 8$
15. $(6 \cdot 15) \cdot \frac{1}{3} = \underline{6\left(15 \cdot \frac{1}{3}\right)}$
16. $(1.5)(2) + 2.8 + 0.2 = (1.5)(2) + \underline{0.2 + 2.8}$
17. $(1 + 18) + 2 = \underline{1 + (18 + 2)}$
18. $(2.25)(-10) = \underline{(-10)(2.25)}$
19. $(44 \cdot 3) \cdot \frac{1}{3} = \underline{(44)\left(3 \cdot \frac{1}{3}\right)}$
20. $(24 + 6)(30) = (\underline{6 + 24})(30)$

EXERCISE 2.2

1. Additive inverse
2. Additive inverse
3. Multiplicative inverse
4. Multiplicative identity
5. Multiplicative identity
6. Multiplicative inverse
7. Additive identity
8. Multiplicative inverse
9. Multiplicative identity
10. Additive inverse
11. $\frac{4}{5} \cdot \frac{5}{4} = \underline{1}$
12. $(7 + 0) \cdot 5 = (\underline{7}) \cdot 5$
13. $\sqrt{19} + -\sqrt{19} = \underline{0}$
14. $x + 0 + 2.6 + 1.4 = \underline{x} + 2.6 + 1.4$
15. $\left(7 \cdot \frac{1}{7}\right)(100) = (\underline{1})(100)$
16. $-\frac{3}{5} + \underline{\frac{3}{5}} = 0$
17. $\left(\frac{9}{10}\right)\left(\underline{\frac{10}{9}}\right) = 1$
18. $\left(-\frac{9}{10}\right)\left(\underline{-\frac{10}{9}}\right) = 1$
19. $(1 \cdot 3)\left(\frac{1}{3}\right) = (\underline{3})\left(\frac{1}{3}\right)$
20. $(\underline{-24} + 24)(30) = (0)(30)$

EXERCISE 2.3

1. $2(8 + 10) = 2 \cdot 8 + \underline{2 \cdot 10}$
2. $4(\underline{7 + 3}) = 4 \cdot 7 + 4 \cdot 3$
3. $(0.25)(1 + 4) = (0.25)(1) + \underline{0.25(4)}$
4. $(5 + 8)20 = 5 \cdot 20 + \underline{8 \cdot 20}$
5. $(15)\left(\frac{2}{3} + \underline{\frac{1}{3}}\right) = 15 \cdot \frac{2}{3} + 15 \cdot \frac{1}{3}$
6. $2(x + 5) = 2 \cdot x + \underline{2 \cdot 5}$
7. $3 \cdot a + 3 \cdot b = 3(\underline{a + b})$
8. $a(\underline{b + c}) = a \cdot b + a \cdot c$
9. $-4 \cdot 9 + -4 \cdot 11 = -4(\underline{9 + 11})$
10. $7 \cdot \frac{1}{2} + 3 \cdot \frac{1}{2} = (\underline{7 + 3})\frac{1}{2}$
11. $2(8 + 10) = 16 + 20 = \underline{36}$
12. $4(7 + 3) = 28 + 12 = \underline{40}$
13. $(0.25)(1 + 4) = 0.25 + 1.00 = \underline{1.25}$
14. $(5 + 8)20 = 100 + 160 = \underline{260}$
15. $(15)\left(\frac{2}{3} + \frac{1}{3}\right) = 10 + 5 = \underline{15}$
16. $0.2(10 + 5) = 2 + 1 = \underline{3}$
17. $\frac{3}{4}\left(\frac{4}{3} + \frac{8}{9}\right) = 1 + \frac{2}{3} = \underline{\frac{5}{3}}$
18. $8(10 + 5) = 80 + 40 = \underline{120}$
19. $(30 + 2)8 = 240 + 16 = \underline{256}$
20. $(7 + 3)\frac{1}{2} = 3.5 + 1.5 = \underline{5}$

EXERCISE 2.4

1. $0 \cdot \frac{7}{8} = \underline{0}$
2. $400(\underline{0}) = 0$
3. $(x)(0) = \underline{0}$
4. $0(5 + x) = \underline{0}$
5. $(15)(\underline{0})(100)(65) = 0$
6. $(a + b)0 = \underline{0}$
7. $(0.85)(10.25)(3.24)(0) = \underline{0}$
8. $(4.5 + 9.9 - 7.5)(\underline{0}) = 0$
9. $(0)(-4 \cdot 9 + 3.5 + 1.2) = \underline{0}$
10. $7 \cdot \frac{1}{2} \cdot x \cdot \frac{2}{7} \cdot 0 \cdot \frac{1}{2} = \underline{0}$

CHAPTER 3

The Number Line and Comparing Numbers

EXERCISE 3.1

1. $-5, -1, 0, 4, 6$

2. $-6, -1, -\frac{1}{2}, 0, 0.\overline{3}$

3. $-4.5, -3, -1\frac{1}{2}, 0.75, 3\frac{1}{4}$

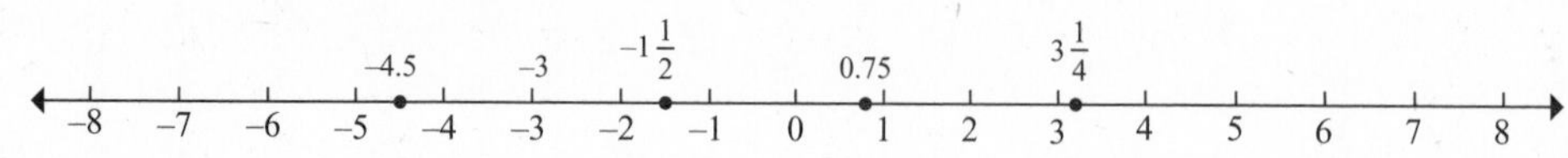

4. $0, 0.5, 2.5, 3.5, 4$

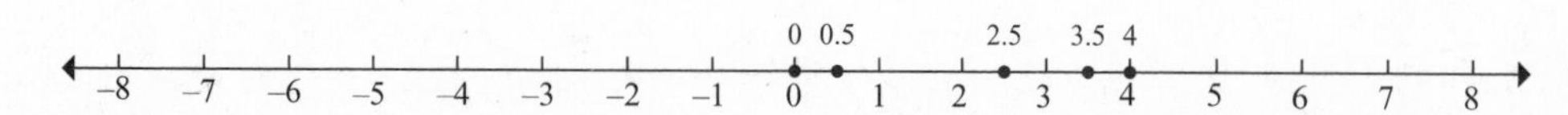

EXERCISE 3.2

1. False
2. True
3. False
4. True
5. False
6. True
7. True
8. False
9. False
10. False
11. False
12. False
13. False
14. True
15. False
16. $-\frac{2}{3}, -0.6, 0.2, \frac{1}{2}, \frac{5}{8}$
17. $-8, -3, -2\frac{1}{2}, -2, 0$
18. $-5, -3, -\frac{2}{3}, 5, 8$
19. $-0.39, -\frac{1}{3}, 0.4, \frac{3}{7}, 1$
20. $-200, -100, -25, 0, 30$

CHAPTER 4

Absolute Value

EXERCISE 4.1

1. 30
2. 0.5
3. $2\frac{1}{3}$
4. 4.8
5. 30
6. 60
7. 140
8. 0
9. -30
10. -30
11. True
12. True
13. False
14. True
15. False
16. False
17. True
18. False
19. False
20. False

EXERCISE 4.2

1. False
2. True
3. True
4. True
5. True
6. False
7. True
8. True
9. True
10. True
11. False
12. True
13. True
14. True
15. True
16. 14
17. −20
18. 3
19. 120
20. 60

CHAPTER 5

Performing Operations with Real Numbers

EXERCISE 5.1

1. $-3 + -6 = \underline{-9}$
2. $11 + 23 = \underline{34}$
3. $-18 + 12 = \underline{-6}$
4. $-100 + 250 = \underline{150}$
5. $-6 + 0 = \underline{-6}$
6. $-78 + 78 = \underline{0}$
7. $-2.5 + 3.25 = \underline{0.75}$
8. $-\frac{1}{2} + \frac{3}{8} = \underline{-\frac{1}{8}}$
9. $-7 - 9 + 25 - 3 = \underline{6}$
10. $0.08 + 2.12 + 0 - 3.2 = \underline{-1}$
11. $25.5 - \frac{3}{4} - 20 = \underline{4.75}$
12. $|100 - 400| = \underline{300}$
13. $|100| + |-400| = \underline{500}$
14. $5 + 8 + 6 - 2 + 7 - 3 = \underline{21}$
15. $12 + 9 - 6 + 6 + 1 - 7 = \underline{15}$
16. 44.95
17. Yes
18. 2,310 feet
19. 3° F
20. 380 miles per hour

EXERCISE 5.2

1. $(-3)(9) = \underline{-27}$
2. $11 \cdot 12 = \underline{132}$
3. $\frac{-18}{-2} = \underline{9}$
4. $-100 \cdot -25 = \underline{2{,}500}$
5. $-16 \cdot 0 = \underline{0}$
6. $\frac{78}{13} = \underline{6}$
7. $(0.25)(-400) = \underline{-100}$
8. $\left(-\frac{1}{2}\right)\left(\frac{4}{3}\right) = -\frac{2}{3}$
9. $(56)\left(-\frac{5}{8}\right) = \underline{-35}$
10. $(-6)(-5)(1.5) = \underline{45}$
11. $\frac{3/5}{-5/6} = \underline{-\frac{1}{2}}$
12. $\left|\frac{-50}{2}\right| = \underline{25}$
13. $\left|\frac{-400}{-100}\right| = \underline{4}$
14. $\frac{0}{89} = \underline{0}$
15. $\frac{99}{0} =$ undefined
16. 36 ounces
17. 10.8°F
18. 13.6 yards
19. −0.625°F
20. 0.498

CHAPTER 6

Exponents

EXERCISE 6.1

1. 3^4
2. $\left(-\frac{1}{2}\right)^7$
3. $(2.5)^2$
4. $(-6)^3$
5. 2^5
6. 16
7. $\frac{1}{81}$
8. 0.64
9. $-\frac{64}{125}$
10. −81
11. 48
12. 25
13. 13
14. −4
15. 0.125
16. 32
17. 100
18. 196
19. $\frac{100}{169}$
20. 6.25

EXERCISE 6.2

1. 1
2. 1
3. 1
4. 1
5. $(3-3)^0 = 0^0 =$ undefined
6. 1
7. 1
8. 1
9. -1
10. 1
11. $\frac{1}{225}$
12. $\frac{16}{9}$
13. 4
14. 25
15. $\frac{81}{625}$
16. -32
17. $\frac{25}{9}$
18. $-\frac{27}{8}$
19. $\frac{7}{3}$
20. $\frac{1}{90{,}000}$

EXERCISE 6.3

1. $100^{1/2}$
2. $625^{1/2}$
3. $(-8)^{1/3}$
4. $27^{4/3}$
5. $16^{5/4}$
6. $1{,}000^{2/3}$
7. $9^{1/4}$
8. $5^{2/2}$
9. $\left(\frac{4}{9}\right)^{3/2}$
10. $(-0.008)^{5/3}$
11. 10
12. 25
13. -2
14. -81
15. 32
16. 100
17. -81
18. 5
19. $\frac{8}{27}$
20. -0.00032

EXERCISE 6.4

1. z^{10}
2. y^3
3. x^{12}
4. $\frac{a^5}{b^5}$
5. $x^8 y^8 z^8$

6. $x^7y^4z^3$

7. $\frac{y^2}{z^3}$

8. x^2yz^4

9. y^5

10. $81x^{12}$

11. $\frac{32a^5}{b^5}$

12. $32x^5y^5z^5$

13. $\frac{x^2}{y^4}$

14. x^2y

15. $\frac{9x^{10}}{y^2}$

16. $\frac{16}{625}$

17. $\frac{5}{3}$

18. $\frac{b^7c^3}{a^7}$

19. $2x^4y^2$

20. $0.3c^4d^5$

CHAPTER 7

Radicals

EXERCISE 7.1

1. $8^{5/3} = \left(\sqrt[3]{8}\right)^5 = 2^5 = 32$

2. $16^{3/4} = \left(\sqrt[4]{16}\right)^3 = 2^3 = 8$

3. $(8x^6y^3)^{5/3} = \left(\sqrt[3]{8x^6y^3}\right)^5 = (2x^2y)^5 = 32x^{10}y^5$

4. $\left(\frac{16z^4}{49x^8y^2}\right)^{1/2} = \sqrt{\frac{16z^4}{49x^8y^2}} = \frac{4z^2}{7x^4y}$

5. $(125r^9s^{21})^{1/3} = \sqrt[3]{125r^9s^{21}} = 5r^3s^7$

6. $(81x^4y^{12}c^{16})^{3/4} = \sqrt[4]{81x^4y^{12}c^{16}} = 3xy^3c^4$

7. $(a^{12})^{-5/3} = \frac{1}{(a^{12})^{5/3}} = \frac{1}{\left(\sqrt[3]{a^{12}}\right)^5} = \frac{1}{(a^4)^5} = \frac{1}{a^{20}}$

8. $(-8a^3b^6c^9)^{1/3} = \sqrt[3]{-8a^3b^6c^9} = -2ab^2c^3$

9. $-(100m^4n^{12})^{1/2} = -\sqrt{100m^4n^{12}} = -10m^2n^6$

10. $-(x^9y^{15})^{4/3} = -\left(\sqrt[3]{x^9y^{15}}\right)^4 = -(x^3y^5)^4 = -x^{12}y^{20}$

11. $(-8x^9y^{15})^{4/3} = \left(\sqrt[3]{-8x^9y^{15}}\right)^4 = (-2x^3y^5)^4 = 16x^{12}y^{20}$

12. $-(625r^4s^8)^{1/4} = -\sqrt[4]{625r^4s^8} = -5rs^2$

13. $(289x^6y^{14})^{1/2} = \sqrt{289x^6y^{14}} = 17x^3y^7$

14. $(169x^2y^2)^{1/2} = \sqrt{169x^2y^2} = 13xy$

15. $(256a^{12}b^8)^{3/4} = \left(\sqrt[4]{256a^{12}b^8}\right)^3 = (4a^3b^2)^3 = 64a^9b^6$

16. $(1{,}000x^3y^6)^{2/3} = \left(\sqrt[3]{1{,}000x^3y^6}\right)^2 = (10xy^2)^2$
$= 100x^2y^4$

17. $(-27x^3y^6z^3)^{2/3} = \left(\sqrt[3]{-27x^3y^6z^3}\right)^2 = (-3xy^2z)^2$
$= 9x^2y^4z^2$

18. $(0.008r^3s^{12})^{1/3} = \sqrt[3]{0.008r^3s^{12}} = 0.2rs^4$

19. $(1.44a^2b^{10}c^4)^{1/2} = \sqrt{1.44a^2b^{10}c^4} = 1.2ab^5c^2$

20. $(0.04x^{40})^{5/2} = \left(\sqrt{0.04x^{40}}\right)^5 = (0.2x^{20})^5$
$= 0.00032x^{100}$

EXERCISE 7.2

1. $4\sqrt{2}$
2. $5\sqrt{3}$
3. $4\sqrt[3]{2}$
4. $10\sqrt{2}$
5. $4\sqrt{11}$
6. $15\sqrt{6}$
7. $-4\sqrt{3}$
8. $4\sqrt{3}$
9. $-2\sqrt{2}$
10. $\sqrt{10}$
11. $7xy^3\sqrt{x}$
12. $8a^4b^4c^5\sqrt{2a}$
13. $2x\sqrt{3}$
14. $6m^2\sqrt[3]{m}$
15. $\frac{1}{2}\sqrt[4]{xy}$
16. $8r^4s\sqrt{5r}$
17. $-21x^2y^3\sqrt{5xy}$
18. $\frac{20a^2}{9b}\sqrt{5ab}$
19. $3x^4y\sqrt[5]{2x^3y^2}$
20. $10x^2y^2\sqrt[3]{x^2}$

EXERCISE 7.3

1. $\frac{\sqrt{2}}{2}$
2. $\frac{\sqrt{3}}{3}$
3. $\frac{1}{2}$
4. $\frac{\sqrt{5}}{5}$
5. $\frac{\sqrt{30}}{6}$
6. $\frac{\sqrt{21}}{7}$
7. $\frac{\sqrt{63}}{9}$
8. $\sqrt{30}$
9. $\frac{\sqrt[3]{48}}{4}$
10. $\sqrt{xy}$

11. $\dfrac{\sqrt{6ab}}{3b}$

12. $\dfrac{x^2\sqrt{6y}}{4y^3}$

13. $-\dfrac{\sqrt{3y}}{3}$

14. $\sqrt{15x}$

15. $\dfrac{3a\sqrt[3]{4abc^2}}{2bc}$

16. $3\sqrt[3]{4x^2y}$

17. $-\sqrt[3]{5m}$

18. $\sqrt[3]{28x^2y^2}$

19. $\dfrac{\sqrt[3]{7}}{8}$

20. $\dfrac{-x\sqrt[3]{36y}}{4y}$

CHAPTER 8

Order of Operations

EXERCISE 8.1

1. $(-5\cdot-4)-(4\cdot-2)=20-(-8)=20+8=28$

2. $\left(\frac{1}{2}\cdot4\right)+(8\cdot3)-(9\cdot5)=2+24-45=-19$

3. $\dfrac{5-16}{-4-7}=\dfrac{-11}{-11}=1$

4. $\dfrac{8+2}{-14+19}+\dfrac{24-36}{-4}=\dfrac{10}{5}+\dfrac{-12}{-4}=2+3=5$

5. $15-|10-24|=15-|-14|=15-14=1$

6. $|20-30|+10=|-10|+10=10+10=20$

7. $\frac{1}{2}(-3-5)=\frac{1}{2}(-8)=-4$

8. $\sqrt{16+9}=\sqrt{25}=5$

9. $7(8-10)=7(-2)=-14$

10. $(-2+5)(8-7)=(3)(1)=3$

11. $6-|5-9|=6-|-4|=6-4=2$

12. $3-(4\cdot8)=3-(32)=-29$

13. $5-2[6-(5\cdot2)]=5-2[6-(10)]=5-2[-4]$
$=5+8=13$

14. $2\sqrt{100-36}=2\sqrt{64}=16$

15. $4-5\sqrt{60+4}=4-5\sqrt{64}=4-40=-36$

16. $\dfrac{4+8}{2-5}+\left(\dfrac{12}{25}\right)\left(\dfrac{75}{6}\right)=\dfrac{12}{-3}+6=-4+6=2$

17. $(5+7)\frac{1}{3}=(12)\frac{1}{3}=4$

18. $2-(4\cdot5)-6(3-4)=2-(20)-6(-1)$
$=2-20+6=-12$

19. $1-[(5\cdot3)-3+2]=1-[15-3+2]$
$=1-14=-13$

20. $10-2[7-(2\cdot3)-(8-3)]=10-2[7-6-(5)]$
$=10-2[-4]=10+8=18$

EXERCISE 8.2

1. $9 - 4(20 - 17)^2 = 9 - 4(3)^2 = 9 - 4(9)$
$= 9 - 36 = -27$

2. $\frac{1}{2} \cdot 4 + 8 \cdot 3 - 9 \cdot 5 = 2 + 24 - 45 = -19$

3. $100 + 8 \cdot 3^2 - 63 + 2(1 + 5) = 100 + 8 \cdot 9 - 63 + 2(6)$
$= 100 + 72 - 63 + 12 = 121$

4. $\frac{-7 + 25}{-3} + |8 - 15| - (5 - 3)^3 = \frac{18}{-3} + |-7| - (2)^3$
$= -6 + 7 - 8 = -7$

5. $15 - \left|\sqrt{100} - 24\right| = 15 - |10 - 24| = 15 - |-14|$
$= 15 - 14 = 1$

6. $|200 - 300| + 10^2 = |-100| + 100$
$= 100 + 100 = 200$

7. $\frac{1}{2}(-3 - 5)^2 = \frac{1}{2}(-8)^2 = \frac{1}{2}(64) = 32$

8. $-2\sqrt{16 + 9} = -2\sqrt{25} = -2 \cdot 5 = -10$

9. $\frac{7(8 - 10)}{14} = \frac{7(-2)}{14} = \frac{-14}{14} = -1$

10. $(-2 + 5)^2 + (8 - 7)^3 = (3)^2 + (1)^3 = 9 + 1 = 10$

11. $\sqrt{36} - \left|5^2 - 9\right| = 6 - |25 - 9| = 6 - |16|$
$= 6 - 16 = -10$

12. $3 \cdot 5 - 4 \cdot 8 = 15 - 32 = -17$

13. $5 - 2[6 - (5 \cdot 2^3 + 4)] = 5 - 2[6 - (5 \cdot 8 + 4)]$
$= 5 - 2[6 - (44)] = 5 - 2[-38] = 5 + 76 = 81$

14. $2(100 - 36)^{1/2} = 2(64)^{1/2} = 2(8) = 16$

15. $2(100^{1/2} - 36^{1/2}) = 2(10 - 6) = 2(4) = 8$

16. $2\left(\frac{5+9}{-2-5}\right) - \left(\frac{1}{3}\right)(14-2) = 2\left(\frac{14}{-2-5}\right) - \left(\frac{1}{3}\right)(14-2)$
$= 2\left(\frac{14}{-7}\right) - \left(\frac{1}{3}\right)(12) = -4 - 4 = -8$

17. $\left(25^{1/2} + 49^{1/2}\right)\frac{1}{3} = (5 + 7)\frac{1}{3} = (12)\frac{1}{3} = 4$

18. $2 - 4 \cdot 5 - 6(5 - 4) = 2 - 9 - 6 = -13$

19. $1 - (5 \cdot 3 - 3 + 2) = 1 - (15 - 3 + 2)$
$= 1 - (14) = -13$

20. $10 - 2[7 - 2 \cdot 3 - (8 - 3)] = 10 - 2[7 - 6 - (5)]$
$= 10 - 2[-4] = 10 + 8 = 18$

CHAPTER 9

Algebraic Expressions and Formulas

EXERCISE 9.1

1. 27
2. 29
3. 54
4. $\frac{5}{7}$
5. -380
6. -280
7. 18
8. 166
9. 400

10. -148

11. 133

12. 6

13. 64

14. 9

15. $\frac{125}{16}$

16. 2

17. -1

18. $\frac{8}{5}$

19. 2

20. $\frac{3}{7}$

EXERCISE 9.2

1. $C = \frac{5}{9}(F - 32) = \frac{5}{9}(212 - 32) = 100$

2. $A = \frac{1}{2}(b_1 + b_2)h = \frac{1}{2}(10 + 6)5 = 40$

3. $d = rt = (65)(3) = 195$

4. $I = \frac{E}{R} = \frac{220}{20} = 11$

5. $A = \frac{1}{2}bh = \frac{1}{2}(14)(12) = 84$

CHAPTER 10

Polynomial Terminology

EXERCISE 10.1

1. Yes

2. Yes

3. Yes

4. Yes

5. Yes

6. No

7. Yes

8. Yes

9. No

10. No

11. (a) -5.25 (b) 4

12. (a) 20 (b) 2

13. (a) 5 (b) 1

14. (a) 13 (b) 6

15. (a) $\sqrt{8}$ (b) 6

16. (a) 1 (b) 4

17. (a) -0.5 (b) 5

18. (a) $-\frac{3}{4}$ (b) 9

19. (a) 50 (b) 6

20. (a) $\frac{2}{3}$ (b) 2

EXERCISE 10.2

1. (a) 4 (b) −7 (c) 3
2. (a) 2 (b) 4 (c) 2
3. (a) 3 (b) 5 (c) 4
4. (a) 5 (b) −5 (c) 5
5. (a) 3 (b) 4 (c) 2
6. (a) 6 (b) 7 (c) 5
7. (a) 4 (b) 1 (c) 3
8. (a) 3 (b) 1 (c) 2
9. (a) 3 (b) 2 (c) 2
10. (a) 1 (b) 20 (c) 1

CHAPTER 11

Adding and Subtracting Polynomials

EXERCISE 11.1

1. $6x + 2x = 8x$
2. $-4x^3 - 3x^3 = -7x^3$
3. $3x - 3x = 0$
4. $-5x^4 + 2x^3$ (These are not like terms, so their sum is only indicated.)
5. $-5z^2 + 4z^2 = -1z^2$
6. $x - x = 0$
7. $-9x^3y^2 - x^3y^2 = -10x^3y^2$
8. $x^4 - 3x^4 = -2x^4$
9. $18 - 20 = -2$
10. $2.5x + 3.5x = 6x$
11. $5x - 9x + 2x = -2x$
12. $-10x^3 - 3x^3 + 14x^3 = x^3$
13. $\frac{1}{2}x - \frac{3}{4}x = -\frac{1}{4}x$
14. $6(x + 5) - 2(x + 5) + 4(x + 5) = 8(x + 5)$
15. $-1.5z^2 + 4.3z^2 + 2.1z^2 = 4.9z^2$
16. $x^2 - 3x^2 + 9x^2 = 7x^2$
17. $-9y - y - 5y + 2y = -13y$
18. $8x - 2y - 5y + 6x = 14x - 7y$
19. $2.1x + 4.0x + 3.5x = 9.6x$
20. $\frac{1}{2}x + \frac{3}{8}y + \frac{1}{4}x + \frac{5}{8}y = \frac{3}{4}x + y$

EXERCISE 11.2

1. $(12x^3 - 5x^2 + 10x - 60) + (3x^3 - 7x^2 - 1)$
   $= 15x^3 - 12x^2 + 10x - 61$
2. $(10x^2 - 5x + 3) + (6x^2 + 5x - 13) = 16x^2 - 10$
3. $(20x^3 - 3x^2 - 2x + 5) + (9x^3 + x^2 + 2x - 15)$
   $= 29x^3 - 2x^2 - 10$
4. $(10x^2 - 5x + 3) - (6x^2 + 5x - 13)$
   $= 4x^2 - 10x + 16$

5. $(20x^3 - 3x^2 - 2x + 5) - (9x^3 + x^2 + 2x - 15)$
$= 11x^3 - 4x^2 - 4x + 20$

6. $(8x^3 - 3x^2 + 6x - 2) + (3x^4 + 2x^3 + x^2 - x)$
$= 3x^4 + 10x^3 - 2x^2 + 5x - 2$

7. $(10y^2 - 15y - 3) + (4y^2 + 5y - 13)$
$= 14y^2 - 10y - 16$

8. $(5x^3 - 4x^2 - 3x + 5) + (6x^3 + 2x^2 - 2x - 15)$
$- (11x^3 - 5x^2 + 2x - 5) = -3x^2 - 7x - 5$

9. $(5x^2 - 10x - 3) - (x^2 - 5x + 10) + (x^3 - 3x^2 - 2x + 1)$
$= x^3 + x^2 - 7x - 12$

10. $(-2x^4 + 3x^3 - 2x + 5) - (7x^3 + x^2 + 2x - 15)$
$- (x^4 - 4) = -3x^4 - 4x^3 - 4x + 24$

11. $2x^2 - 3x + 5 + 4x^2 + 6x - 3 = 6x^2 + 3x + 2$

12. $0 - (3a^3 - 4a + 7) = -3a^3 + 4a - 7$

13. $z^3 + 2z^2 - z - (2z^2 + z - 1) = z^3 - 2z + 1$

14. $a^2 - (a^2 - 4a + 6) = 4a - 6$

15. $x^3 + 2x^2 - 2x^2 = x^3$

16. $(x^3 - 3x^2) - (5x^3 - 4x + 8) = -4x^3 - 3x^2 + 4x - 8.$

17. $(5x^4 - 2x^3) - (3x^2 - 3x + 1) = 5x^4 - 2x^3$
$- 3x^2 + 3x - 1$

18. $(2x^2 + 4xy + y^2) + (x^2 - y^2) + (2y^2 - 4xy - x^2)$
$= 2x^2 + 2y^2$

19. $(x^2 - 2xy + y^2 + x^2 + 2xy + y^2) - (x^2 - 4xy + 4y^2)$
$= x^2 + 4xy - 2y^2$

20. $180 - \left(5x + 15 + 4x - 10\right) = 180 - 9x - 5$
$= 175 - 9x$, in degrees

CHAPTER 12

Multiplying Polynomials

EXERCISE 12.1

1. $(-4x^3)(3x^2) = -12x^5$

2. $(2x^3y^5)(6xy^2) = 12x^4y^7$

3. $(-5x)(-2x) = 10x^2$

4. $(5x)(-4) = -20x$

5. $(-2x^2y^3)(-5xy^3)(-xy) = -10x^4y^7$

6. $(-4y^3)(-z) = 4y^3z$

7. $\left(\frac{1}{2}m^4n^2\right)(8m^2n) = 4m^6n^3$

8. $(-5xyz)(2xy^2) = -10x^2y^3z$

9. $(xy^2)(-2x)(-4x^2y)(3xy) = 24x^5y^4$

10. $(-2a)(-2a)(-2a)(-2a)(-2a) = -32a^5$

11. $5xy \cdot 3 = 15xy$

12. $9ab^4(-1) = -9ab^4$

13. $(-6x^5y^4z)(2x^3z^2) = -12x^8y^4z^3$

14. $10x \cdot 6x = 60x^2$, in feet^2

15. $7x \cdot 3 = 21x$, in miles

EXERCISE 12.2

1. $5(x+3) = 5x + 15$
2. $x(4x-5) = 4x^2 - 5x$
3. $-2(2x+3) = -4x - 6$
4. $3x(2x-1) = 6x^2 - 3x$
5. $a(c+d) = ac + ad$
6. $2x(x-5) = 2x^2 - 10x$
7. $5x(2x-3) = 10x^2 - 15x$
8. $4(2x-3) = 8x - 12$
9. $-5x^3y^2(2x^2 - 6xy^2 + 3) = -10x^5y^2 + 30x^4y^4 - 15x^3y^2$
10. $z^3(3x^4z + 4x^3z^2 - 3xz + 5) = 3x^4z^4 + 4x^3z^5 - 3xz^4 + 5z^3$
11. $(x+3)x = x^2 + 3x$
12. $(9a+5b)(-1) = -9a - 5b$
13. $2z(-5z^2 + 3z - 2) = -10z^3 + 6z^2 - 4z$
14. $(2x-3)(5x) = 10x^2 - 15x$, in meter^2
15. $x(2x+7) = (2x^2 + 7x)$ dollars

EXERCISE 12.3

1. $(x+3)(x-2) = x^2 + x - 6$
2. $(4x-3)(4x+3) = 16x^2 - 9$
3. $(2x-y)(x+2y) = 2x^2 + 3xy - 2y^2$
4. $(5x+4)(2x-3) = 10x^2 - 7x - 12$
5. $(z-2)(z+5) = z^2 + 3z - 10$
6. $(x+y)^2 = x^2 + 2xy + y^2$
7. $(a-b)^2 = a^2 - 2ab + b^2$
8. $\left(x - \sqrt{2}\right)\left(x + \sqrt{2}\right) = x^2 - 2$
9. $\left(\sqrt{7} - \sqrt{3}\right)\left(\sqrt{7} + \sqrt{3}\right) = 7 - 3 = 4$
10. $(x^2 - 5)(x^2 + 5) = x^4 - 25$
11. $(2x+3)(x-1) = 2x^2 + x - 3$
12. $(6+y)(5-2y) = 30 - 7y - 2y^2$
13. $(x+a)(x-a) = x^2 - a^2$
14. $(x+5)(3x-1) = 3x^2 + 14x - 5$, in feet^2
15. $\frac{1}{2}(3x+4)(2x+5) = \frac{1}{2}(6x^2 + 23x + 20)$, in meter^2

EXERCISE 12.4

1. $(x+3)(x^2 - 6x + 9) = x^3 - 6x^2 + 9x + 3x^2 - 18x + 27 = x^3 - 3x^2 - 9x + 27$
2. $(2z^2 - z - 3)(4z^2 + 2z - 5) = 8z^4 + 4z^3 - 10z^2 - 4z^3 - 2z^2 + 5z - 12z^2 - 6z + 15 = 8z^4 - 24z^2 - z + 15$
3. $(ax+b)(cx+d) = acx^2 + adx + bcx + bd$
4. $(5x-4)(2x-3) = 10x^2 - 23x + 12$
5. $(3x^2 + 2x - 7)(x - 8) = 3x^3 - 24x^2 + 2x^2 - 16x - 7x + 56 = 3x^3 - 22x^2 - 23x + 56$
6. $(3m^2 - 4n^2)(2m^2 - 3n^2) = 6m^4 - 9m^2n^2 - 8m^2n^2 + 12n^4 = 6m^4 - 17m^2n^2 + 12n^4$
7. $(a-b)(a^2 + ab + b^2) = a^3 + a^2b + ab^2 - a^2b - ab^2 - b^3 = a^3 - b^3$
8. $(z^3 - z^2 + z - 1)(z^2 - z + 1) = z^5 - z^4 + z^3 - z^4 + z^3 - z^2 + z^3 - z^2 + z - z^2 + z - 1 = z^5 - 2z^4 + 3z^3 - 3z^2 + 2z - 1$
9. $(x^3 - x^2 - x)(x^2 + 2x - 3) = x^5 + 2x^4 - 3x^3 - x^4 - 2x^3 + 3x^2 - x^3 - 2x^2 + 3x = x^5 + x^4 - 6x^3 + x^2 + 3x$

10. $(2y^2 - 5y + 3)(-3y - 4) = -6y^3 - 8y^2 + 15y^2 + 20y - 9y - 12 = -6y^3 + 7y^2 + 11y - 12$

11. $\left(a + b - \sqrt{2}c\right)\left(a + b + \sqrt{2}c\right) = a^2 + ab + \sqrt{2}ac + ab + b^2 + \sqrt{2}bc - \sqrt{2}ac - \sqrt{2}bc - 2c^2 = a^2 + 2ab + b^2 - 2c^2$

12. $(3z - 1)(z^4 + 2z^3 - 3z^2 + 7z + 5) = 3z^5 + 6z^4 - 9z^3 + 21z^2 + 15z - z^4 - 2z^3 + 3z^2 - 7z - 5 = 3z^5 + 5z^4 - 11z^3 + 24z^2 + 8z - 5$

13. $(3m^2 + 5n^2)(2m^2 - 7n^2) = 6m^4 - 11m^2n^2 - 35n^4$

14. $(x - y - z)(x + y - z) = x^2 + xy - xz - xy - y^2 + yz - xz - yz + z^2 = x^2 - 2xz - y^2 + z^2$

15. $5x(x + 4)(2x - 1) = 5x(2x^2 + 7x - 4) = 10x^3 + 35x^2 - 20x$

EXERCISE 12.5

1. $(a + 6)(a - 6) = a^2 - 36$
2. $(z + 3)^2 = z^2 + 6z + 9$
3. $(x - 2)^2 = x^2 - 4x + 4$
4. $(x + 2)(x^2 - 2x + 4) = x^3 + 8$
5. $(x - 2)(x^2 + 2x + 4) = x^3 - 8$
6. $(x + 3)^3 = x^3 + 9x^2 + 27x + 27$
7. $(a - b)(a^2 + ab + b^2) = a^3 - b^3$
8. $(z - 2)^3 = z^3 - 6z^2 + 12x - 8$
9. $(2m - 1)^2 = 4m^2 - 4m + 1$
10. $(a - 1)(a^2 + a + 1) = a^3 - 1$
11. $\left(x - \sqrt{3}\right)\left(x + \sqrt{3}\right) = x^2 - 3$
12. $(2a + 3)^2 = 4a^2 + 12a + 9$
13. $(m + 1)^3 = m^3 + 3m^2 + 3m + 1$
14. $(2x - 1)(4x^2 + 2x + 1) = 8x^3 - 1$
15. $(x + 4)^2 = x^2 + 8x + 16$, in yd^2

CHAPTER 13

Simplifying Polynomial Expressions

EXERCISE 13.1

1. $6x + (5y + 10) = 6x + 5y + 10$
2. $9 + (-5y + 4) = 9 - 5y + 4 = 13 - 5y$
3. $-2a^2 - (-a^2 + 4a) = -2a^2 + a^2 - 4a = -a^2 - 4a$
4. $2z^2 - 4 - (-5 + 7z^2) = 2z^2 - 4 + 5 - 7z^2 = -5z^2 + 1$
5. $8a^4 - (2a^4 - 5) + (2a^2 - 1) - 6 = 8a^4 - 2a^4 + 5 + 2a^2 - 1 - 6 = 6a^4 + 2a^2 - 2$

6. $4x+[3-(2x-5)]=4x+3-2x+5=2x+8$

7. $4x-[3-(2x-5)]=4x-[3-2x+5]$
$=4x-[-2x+8]=4x+2x-8=6x-8$

8. $3x-4y+[2x-(3x-4y)]-(5x-7y)$
$=3x-4y+[2x-3x+4y]-(5x-7y)$
$=3x-4y-x+4y-5x+7y=-3x+7y$

9. $=m^2-[m+(2m^2-1)]+3m+[2m-(m^2-1)]-3$
$=m^2-[m+2m^2-1]+3m+[2m-m^2+1]-3$
$=m^2-m-2m^2+1+3m+2m-m^2+1-3$
$=-2m^2+4m-1$

10. $a-(a^2-(a+1))+a^2=a-(a^2-a-1)+a^2$
$=a-a^2+a+1+a^2=2a+1$

EXERCISE 13.2

1. $6x+2(5x-3)=6x+10x-6=16x-6$

2. $9-1(-5z+4)+3z=9+5z-4+3z$
$=8z+5$

3. $3a^2-(-a^2+4a+1)+2a-a(a+3)-8$
$=3a^2+a^2-4a-1+2a-a^2-3a-8$
$=3a^2-5a-9$

4. $8z^2-4-2(-5+7z^2)=8z^2-4+10-14z^2$
$=-6z^2+6$

5. $(x+2)(x-2)+(x+2)^2=x^2-4+x^2$
$+4x+4=2x^2+4x$

6. $(3m-5)(4m-1)+(m-3)(m+4)$
$=12m^2-23m+5+m^2+m-12$
$=13m^2-22m-7$

7. $4x+2[3x-2(2x-5)]=4x+2[3x-4x+10]$
$=4x+2[-x+10]=4x-2x+20=2x+20$

8. $3x^2-4y^2-x[2x-(3x-4)]-y(5-7y)$
$=3x^2-4y^2-x[-x+4]-y(5-7y)$
$=3x^2-4y^2+x^2-4x-5y+7y^2$
$=4x^2-4x+3y^2-5y$

9. $3x(x^2-9)-(x-3)(x^2+3x+9)$
$=3x^3-27x-(x^3-27)$
$=3x^3-27x-x^3+27$
$=2x^3-27x+27$

10. $a^2-a-4(2a-(a+1)+1)+3a(a-1)$
$=a^2-a-4(a)+3a(a-1)$
$=a^2-a-4a+3a^2-3a=4a^2-8a$

CHAPTER 14

Dividing Polynomials

EXERCISE 14.1

1. $\dfrac{4x^4y-8x^3y^3+16xy^4}{4xy}=x^3-2x^2y^2+4y^3$

2. $\dfrac{16x^5y^2}{16x^5y^2}=1$

3. $\dfrac{15x^5-30x^2}{-5x}=-3x^4+6x$

4. $\dfrac{-14x^4+21x^2}{-7x^2}=2x^2-3$

5. $\dfrac{25x^4y^2}{-5x}=-5x^3y^2$

6. $\dfrac{6x^5y^2-8x^3y^3+10xy^6}{2xy^2}=3x^4-4x^2y+5y^4$

7. $\dfrac{-10x^4y^4z^4 - 20x^2y^5z^2}{10x^2y^3z} = -x^2yz^3 - 2y^2z$

8. $\dfrac{-1.8x^5}{0.3x} = -6x^4$

9. $\dfrac{-18x^5 + 5}{3x^5} = -6 + \dfrac{5}{3x^5}$

10. $\dfrac{7a^6b^3 - 14a^5b^2 - 42a^4b^2 + 7a^3b^2}{7a^3b^2}$
$= a^3b - 2a^2 - 6a + 1$

11. $\dfrac{14x^5y}{-xy} = -14x^4$

12. $\dfrac{9ab^4}{-3a^2b} = -\dfrac{3b^3}{a}$

13. $\dfrac{3x^2 - 15x}{3x} = x - 5$

14. $\dfrac{24m^2}{6m} = 4m$

15. $\dfrac{25x}{2} = 12.5x$, in miles per hour

EXERCISE 14.2

1.
$$\begin{array}{r} x - 2 \\ x + 3\overline{\smash{\big)}\,x^2 + x - 4} \\ \underline{x^2 + 3x\phantom{{}-4}} \\ -2x - 4 \\ \underline{-2x - 6} \\ 2 \end{array}$$

Answer: $x - 2 + \dfrac{2}{x + 3}$

2.
$$\begin{array}{r} x - 4 \\ x - 5\overline{\smash{\big)}\,x^2 - 9x + 20} \\ \underline{x^2 - 5x\phantom{{}+20}} \\ -4x + 20 \\ \underline{-4x + 20} \\ 0 \end{array}$$

Answer: $x - 4$

3.
$$\begin{array}{r} 2x^2 + 9x + 23 \\ x - 4\overline{\smash{\big)}\,2x^3 + x^2 - 13x + 6} \\ \underline{2x^3 - 8x^2\phantom{{}-13x+6}} \\ 9x^2 - 13x + 6 \\ \underline{9x^2 - 36x\phantom{{}+6}} \\ 23x + 6 \\ \underline{23x - 92} \\ 98 \end{array}$$

Answer: $2x^2 + 9x + 23 + \dfrac{98}{x - 4}$

4.
$$\begin{array}{r} x^3 + 8 \\ x^3 - 8\overline{\smash{\big)}\,x^6 + 0 + 0 + 0 + 0 + 0 - 64} \\ \underline{x^6 - 8x^3\phantom{{}+0+0+0-64}} \\ 8x^3 - 64 \\ \underline{8x^3 - 64} \\ 0 \end{array}$$

Answer: $x^3 + 8$

5.
$$\begin{array}{r} x + 4 \\ x - 3\overline{\smash{\big)}\,x^2 + x - 12} \\ \underline{x^2 - 3x\phantom{{}-12}} \\ 4x - 12 \\ \underline{4x - 12} \\ 0 \end{array}$$

Answer: $x + 4$

CHAPTER 15

Factoring Polynomials

EXERCISE 15.1

1. $4x + 4y = 4(x + y)$
2. $3x + 6 = 3(x + 2)$
3. $12x^8y^3 - 8x^6y^7z^2 = 4x^6y^3(3x^2 - 2y^4z^2)$
4. $15x^2 - 3x = 3x(5x - 1)$
5. $x^3y - xy + y = y(x^3 - x + 1)$
6. $\frac{1}{2}ax - \frac{1}{2}ay = \frac{1}{2}a(x - y)$
7. $x(w - z) - y(w - z) = (w - z)(x - y)$
8. $1.5a^2b + 4.5ab + 7.5ab^2 = 1.5ab(a + 3 + 5b)$
9. $mx + my + 5x + 5y = m(x + y) + 5(x + y)$
   $= (x + y)(m + 5)$
10. $xy + xy^2 + xy^3 + xy^4 = xy(1 + y + y^2 + y^3)$
11. $-5x - 5y = -5(x + y)$
12. $-4x + 8 = -4(x - 2)$
13. $-24x^8y^3 - 8x^6y^7z^2 = -8x^6y^3(3x^2 + y^4z^2)$
14. $15x^2 - 3x = -3x(-5x + 1)$
15. $ar - rt - r = -r(-a + t + 1)$

EXERCISE 15.2

1. $16x^2 - 36 = 4(x^2 - 9) = 4(x + 3)(x - 3)$
2. $x^2 - y^2 = (x + y)(x - y)$
3. $36x^2 - 49 = (6x + 7)(6x - 7)$
4. $x^2 - 1 = (x + 1)(x - 1)$
5. $\frac{1}{4}x^2 - 25 = \left(\frac{1}{2}x + 5\right)\left(\frac{1}{2}x - 5\right)$
6. $z^2 - 0.36 = (z + 0.6)(z - 0.6)$
7. $100x^2y^2 - 81z^2 = (10xy + 9z)(10xy - 9z)$
8. $a^2 - \frac{1}{49} = \left(a + \frac{1}{7}\right)\left(a - \frac{1}{7}\right)$
9. $x^2 - 3 = \left(x + \sqrt{3}\right)\left(x - \sqrt{3}\right)$
10. $z^2 - 5 = \left(z + \sqrt{5}\right)\left(z - \sqrt{5}\right)$
11. $121 - 25c^2 = (11 + 5c)(11 - 5c)$
12. $x^4 - y^2 = (x^2 + y)(x^2 - y)$
13. $x^2y^2 - z^2 = (xy + z)(xy - z)$
14. $x^4y^2 - 1 = (x^2y + 1)(x^2y - 1)$
15. $0.64 - x^2 = (0.8 + x)(0.8 - x)$

EXERCISE 15.3

1. $x^2 - 4x + 4 = (x - 2)^2$
2. $x^2 + 10xy + 25y^2 = (x + 5y)^2$
3. $36x^2 + 12x + 1 = (6x + 1)^2$
4. $4x^2 - 12x + 9 = (2x - 3)^2$
5. $49a^2 + 56a + 16 = (7a + 4)^2$
6. $x^2 + x + \frac{1}{4} = \left(x + \frac{1}{2}\right)^2$
7. $100 + 140x + 49x^2 = (10 + 7x)^2$
8. $36x^2 + 60x + 25 = (6x + 5)^2$
9. $81 - 36x + 4x^2 = (9 - 2x)^2$
10. $1 - 2x + x^2 = (1 - x)^2$

EXERCISE 15.4

1. $x^3 + 125 = (x + 5)(x^2 - 5x + 25)$
2. $y^3 - 27 = (y - 3)(y^2 + 3y + 9)$
3. $64a^3 + 1 = (4a + 1)(16a^2 - 4a + 1)$
4. $8z^3 - 125 = (2z - 5)(4z^2 + 10z + 25)$
5. $125a^3 + 27 = (5a + 3)(25a^2 - 15a + 9)$
6. $8x^3 - 125 = (2x - 5)(4x^2 + 10x + 25)$
7. $64y^3 - 125 = (4y - 5)(16y^2 + 20y + 25)$
8. $216x^3 + 1{,}000 = 8(3x + 5)(9x^2 - 15x + 25)$
9. $8x^3 + 125y^3z^6 = (2x + 5yz^2)(4x^2 - 10xyz^2 + 25y^2z^4)$
10. $125a^6 - 27 = (5a^2 - 3)(25a^4 + 15a^2 + 9)$

EXERCISE 15.5

1. $x^2 + 5x + 6 = (x + 2)(x + 3)$
2. $x^2 + 5x - 6 = (x + 6)(x - 1)$
3. $y^2 + 9y - 10 = (y + 10)(y - 1)$
4. $b^2 + 7b - 98 = (b + 14)(b - 7)$
5. $z^2 - 5z - 14 = (z + 2)(z - 7)$
6. $2x^2 + 5x - 3 = (2x - 1)(x + 3)$
7. $6x^2 - x - 1 = (3x + 1)(2x - 1)$
8. $9y^2 + 9y - 4 = (3y + 4)(3y - 1)$
9. $49b^2 - 21b - 10 = (7b + 2)(7b - 5)$
10. $16z^2 - 16z - 5 = (4z - 5)(4z + 1)$

CHAPTER 16

Fundamental Concepts of Rational Expressions

EXERCISE 16.1

1. $x \neq -4$

2. No restriction

3. $x \neq 1$

4. $x \neq -2, x \neq -3$

5. $x \neq 0$

6. No restriction

7. $x \neq 2, x \neq -2$

8. No restriction

9. $x \neq 3, x \neq -3$

10. $x \neq -\frac{5}{2}$

EXERCISE 16.2

1. $\frac{-8x^2y}{16xy^2} = \frac{-x}{2y}$

2. $\frac{24a^3x^2}{32ay} = \frac{3a^2x^2}{4y}$

3. $\frac{18x^4}{30x} = \frac{3x^3}{5}$

4. $\frac{10x^2z^3}{xz^4} = \frac{10x}{z}$

5. $\frac{5x^2(x+3)}{15x(x+3)} = \frac{x}{3}$

6. $\frac{12xy^2}{6x^4(x-y)} = \frac{2y^2}{x^3(x-y)}$

7. $\frac{xy - x^2y^2}{xz - x^2yz} = \frac{xy(1-xy)}{xz(1-xy)} = \frac{y}{z}$

8. $\frac{4/3\,\pi r^3}{4\pi r^2} = \frac{r}{3}$

9. $\frac{4x(y-2z)}{2x^4(y-2z)} = \frac{2}{x^3}$

10. $\frac{10a^2b^5}{5b^2(a+b)} = \frac{2a^2b^3}{(a+b)}$

11. $\frac{x^2-16}{x^2-8x+16} = \frac{(x+4)(x-4)}{(x-4)^2} = \frac{(x+4)}{(x-4)}$

12. $\frac{z^2+4z-5}{z^2+8z+15} = \frac{(z+5)(z-1)}{(z+3)(z+5)} = \frac{(z-1)}{(z+3)}$

13. $\frac{2y^2+4y-30}{3y^2+21y+30} = \frac{2(y^2+2y-15)}{3(y^2+7y+10)}$

$= \frac{2(y+5)(y-3)}{3(y+5)(y+2)} = \frac{2(y-3)}{3(y+2)}$

14. $\frac{3xy^3-27xy}{6xy^2+6xy-72x} = \frac{3xy(y^2-9)}{6x(y^2+y-12)}$

$= \frac{y(y+3)(y-3)}{2(y+4)(y-3)} = \frac{y(y+3)}{2(y+4)}$

15. $\frac{a^2-12a+36}{a^2-3a-18} = \frac{(a-6)^2}{(a-6)(a+3)} = \frac{(a-6)}{(a+3)}$

16. $\frac{4z^2+16z+16}{6z^2+18z+12} = \frac{4(z^2+4z+4)}{6(z^2+3z+2)}$

$= \frac{2(z+2)^2}{3(z+2)(z+1)} = \frac{2(z+2)}{3(z+1)}$

17. $\dfrac{x^3 - xy^2}{xy(x^2 - 2xy + y^2)} = \dfrac{x(x^2 - y^2)}{xy(x^2 - 2xy + y^2)}$
$= \dfrac{(x + y)(x - y)}{y(x - y)^2} = \dfrac{(x + y)}{y(x - y)}$

18. $\dfrac{3t + 15}{t^2 - 25} = \dfrac{3(t + 5)}{(t + 5)(t - 5)} = \dfrac{3}{(t - 5)}$

19. $\dfrac{2x - 10}{x^2 - 10x + 25} = \dfrac{2(x - 5)}{(x - 5)^2} = \dfrac{2}{(x - 5)}$

20. $\dfrac{(x-2)}{(x^3-8)} = \dfrac{(x-2)}{(x-2)(x^2+2x+4)} = \dfrac{1}{(x^2+2x+4)}$

EXERCISE 16.3

1. $\dfrac{x}{2y} = \dfrac{x \cdot 8xy}{2y \cdot 8xy} = \dfrac{8x^2y}{16xy^2}$

2. $\dfrac{2y}{3x^2} = \dfrac{2y \cdot 4x^3y}{3x^2 \cdot 4x^3y} = \dfrac{8x^3y^2}{12x^5y}$

3. $\dfrac{3x^3}{5} = \dfrac{3x^3 \cdot 6x}{5 \cdot 6x} = \dfrac{18x^4}{30x}$

4. $\dfrac{10xz}{1} = \dfrac{10xz \cdot xz^4}{1 \cdot xz^4} = \dfrac{10x^2z^5}{xz^4}$

5. $\dfrac{1}{x^3} = \dfrac{1 \cdot x^5}{x^3 \cdot x^5} = \dfrac{x^5}{x^8}$

6. $\dfrac{x}{3xy} = \dfrac{x \cdot 6}{3xy \cdot 6} = \dfrac{6x}{18xy}$

7. $\dfrac{1}{x + 3} = \dfrac{1(x + 3)}{(x + 3)(x + 3)} = \dfrac{(x + 3)}{(x + 3)^2}$

8. $\dfrac{x + 3}{x - 3} = \dfrac{(x + 3)(x + 3)}{(x - 3)(x + 3)} = \dfrac{(x + 3)^2}{(x + 3)(x - 3)}$

9. $\dfrac{3x(x + 2)}{(x - 2)} = \dfrac{3x(x + 2) \cdot (x - 4)}{(x - 2) \cdot (x - 4)}$
$= \dfrac{3x(x + 2)(x - 4)}{(x - 2)(x - 4)}$

10. $\dfrac{2a^2}{b^2} = \dfrac{2a^2 \cdot 5(a + b)}{b^2 \cdot 5(a + b)} = \dfrac{10a^2(a + b)}{5b^2(a + b)}$

11. $\dfrac{x + 4}{x - 4} = \dfrac{(x + 4) \cdot (x - 4)}{(x - 4) \cdot (x - 4)} = \dfrac{(x + 4)(x - 4)}{(x - 4)^2}$

12. $\dfrac{z + 2}{z + 5} = \dfrac{(z + 2)(z + 3)}{(z + 5)(z + 3)}$

13. $\dfrac{2y - 6}{3y + 6} = \dfrac{2(y - 3) \cdot (y + 5)}{3(y + 2) \cdot (y + 5)} = \dfrac{2(y - 3)(y + 5)}{3(y + 2)(y + 5)}$

14. $2x^2y = \dfrac{2x^2y \cdot 2x}{1 \cdot 2x} = \dfrac{4x^3y}{2x}$

15. $\dfrac{x + y}{2xy^2} = \dfrac{(x + y) \cdot 3x}{2xy^2 \cdot 3x} = \dfrac{3x(x + y)}{6x^2y^2}$

16. $\dfrac{5}{x + 3} = \dfrac{5(x - 2)}{(x + 3)(x - 2)}$

17. $\dfrac{2a}{a + 5} = \dfrac{2a(a - 3)}{(a + 5)(a - 3)}$

18. $\dfrac{4c}{c - 2} = \dfrac{4c(c - 2)}{(c - 2)(c - 2)} = \dfrac{4c(c - 2)}{(c - 2)^2}$

19. $\dfrac{x}{3(x + 4)} = \dfrac{x \cdot 2(4x + 1)}{3(x + 4) \cdot 2(4x + 1)}$
$= \dfrac{2x(4x + 1)}{6(x + 4)(4x + 1)}$

20. $\dfrac{5x}{(x + 4)(x + 3)} = \dfrac{5x(x - 1)}{(x + 4)(x + 3)(x - 1)}$

CHAPTER 17

Multiplying and Dividing Rational Expressions

EXERCISE 17.1

1. $\frac{2a}{3b} \cdot \frac{6b}{a} = 4$

2. $\frac{24x^2}{48y^2} \cdot y^2 = \frac{x^2}{2}$

3. $\frac{4x^4}{5} \cdot \frac{10}{x^3} = 8x$

4. $\frac{27x^2y^3}{8z^3} \cdot \frac{16x^2}{9xy} = \frac{6x^3y^2}{z^3}$

5. $\frac{2a^2x^4}{9ab^2} \cdot \frac{3x^2b^3}{8ax^3} \cdot \frac{6bx^2}{7a^5b} = \frac{bx^5}{14a^5}$

6. $\frac{a+b}{a-b} \cdot \frac{a-b}{a+b} = 1$

7. $\frac{x^2-5x-6}{x^2+6x+8} \cdot \frac{x^2+x-2}{x^2-2x-3}$
$= \frac{(x-6)(x+1)}{(x+2)(x+4)} \cdot \frac{(x+2)(x-1)}{(x-3)(x+1)}$
$= \frac{(x-6)(x-1)}{(x+4)(x-3)}$

8. $\frac{x+y}{3xy} \cdot \frac{6xy}{(x+y)^2} = \frac{2}{(x+y)}$

9. $\frac{a+6}{a-3} \cdot (a-3) = a+6$

10. $\frac{(x-2)^5}{4x} \cdot \frac{12x^3}{(x-2)^3} = 3x^2(x-2)^2$

11. $\frac{(x+y)(y+2)}{(x+2)} \cdot \frac{(x+2)(x+2y)}{2(y+2)}$
$= \frac{(x+y)(x+2y)}{2}$

12. $\frac{(x-2)(y+2)}{(x+2)(y-2)} \cdot \frac{(x+2)(y-2)}{(x-2)(y-2)} = \frac{(y+2)}{(y-2)}$

13. $\frac{4x^2+10}{x-3} \cdot \frac{x^2-9}{6x^2+15}$
$= \frac{2(2x^2+5)}{(x-3)} \cdot \frac{(x+3)(x-3)}{3(2x^2+5)}$
$= \frac{2(x+3)}{3}$

14. $\frac{4x-4}{x^2-y^2} \cdot \frac{x^3y^2-x^2y^3}{2x-2}$
$= \frac{4(x-1)}{(x+y)(x-y)} \cdot \frac{x^2y^2(x-y)}{2(x-1)}$
$= \frac{2x^2y^2}{(x+y)}$

15. $\frac{a^2-12a+36}{3a^3} \cdot \frac{6a}{a^2-3a-18}$
$= \frac{(a-6)^2}{3a^3} \cdot \frac{6a}{(a-6)(a+3)}$
$= \frac{2(a-6)}{a^2(a+3)}$

16. $\frac{3}{6z^2+18z+12} \cdot \frac{4z^2+16z+16}{4}$
$= \frac{3}{6(z^2+3z+2)} \cdot \frac{4(z^2+4z+4)}{4}$
$= \frac{1}{2(z+1)(z+2)} \cdot \frac{(z+2)^2}{1} = \frac{(z+2)}{2(z+1)}$

17. $\frac{1}{(x^2-2xy+y^2)} \cdot \frac{x^3-xy^2}{xy}$
$= \frac{1}{(x-y)^2} \cdot \frac{x(x^2-y^2)}{xy}$
$= \frac{1}{(x-y)^2} \cdot \frac{x(x+y)(x-y)}{xy} = \frac{(x+y)}{y(x-y)}$

18. $\dfrac{t-5}{3t^2-75}\cdot\dfrac{6t+30}{2t}=\dfrac{(t-5)}{3(t^2-25)}\cdot\dfrac{6(t+5)}{2t}$

$=\dfrac{(t-5)}{3(t+5)(t-5)}\cdot\dfrac{6(t+5)}{2t}=\dfrac{1}{t}$

19. $\dfrac{x^2-7x+12}{x^2-x-6}\cdot\dfrac{x^2+7x+10}{x^2+x-20}$

$=\dfrac{(x-3)(x-4)}{(x-3)(x+2)}\cdot\dfrac{(x+2)(x+5)}{(x+5)(x-4)}=1$

20. $\dfrac{x^2-13x+42}{x^2+2x}\cdot\dfrac{x^2+x-2}{2x^2-14x}$

$=\dfrac{(x-7)(x-6)}{x(x+2)}\cdot\dfrac{(x+2)(x-1)}{2x(x-7)}$

$=\dfrac{(x-6)(x-1)}{2x^2}$

EXERCISE 17.2

1. $\dfrac{3}{x}\div\dfrac{5}{x}=\dfrac{3}{x}\cdot\dfrac{x}{5}=\dfrac{3}{5}$

2. $\dfrac{4xy}{3y}\div\dfrac{2x}{y}=\dfrac{4xy}{3y}\cdot\dfrac{y}{2x}=\dfrac{2y}{3}$

3. $\dfrac{3x^2y}{8z^2}\div\dfrac{6xy^3}{z}=\dfrac{3x^2y}{8z^2}\cdot\dfrac{z}{6xy^3}=\dfrac{x}{16y^2z}$

4. $5x^2y\div\dfrac{10xy^2}{3}=\dfrac{5x^2y}{1}\cdot\dfrac{3}{10xy^2}=\dfrac{3x}{2y}$

5. $\dfrac{2m^3n}{11}\div\dfrac{4mn^3}{33}=\dfrac{2m^3n}{11}\cdot\dfrac{33}{4mn^3}=\dfrac{3m^2}{2n^2}$

6. $\dfrac{18}{x}\div 6=\dfrac{18}{x}\cdot\dfrac{1}{6}=\dfrac{3}{x}$

7. $\dfrac{2x^4y^2}{5z}\div 8xy^2=\dfrac{2x^4y^2}{5z}\cdot\dfrac{1}{8xy^2}=\dfrac{x^3}{20z}$

8. ${}^4\!/_3\,\pi r^3\div 4\pi r^2=\dfrac{4\pi r^3}{3}\cdot\dfrac{1}{4\pi r^2}=\dfrac{r}{3}$

9. $4xyz\div\dfrac{2x^2y}{3z^2}=4xyz\cdot\dfrac{3z^2}{2x^2y}=\dfrac{6z^3}{x}$

10. $\dfrac{9a^2b^4}{10c}\div 27abc=\dfrac{9a^2b^4}{10c}\cdot\dfrac{1}{27abc}=\dfrac{ab^3}{30c^2}$

11. $\dfrac{4x-4y}{5x+5y}\div\dfrac{4}{25}=\dfrac{4x-4y}{5x+5y}\cdot\dfrac{25}{4}=\dfrac{4(x-y)}{5(x+y)}\cdot\dfrac{25}{4}$

$=\dfrac{5(x-y)}{(x+y)}$

12. $\dfrac{4x-4}{x^2-16}\div\dfrac{x-1}{x-4}=\dfrac{4(x-1)}{(x+4)(x-4)}\cdot\dfrac{(x-4)}{(x-1)}$

$=\dfrac{4}{x+4}$

13. $\dfrac{x^2-y^2}{15(x+y)^2}\div\dfrac{x-y}{5x+5y}=\dfrac{x^2-y^2}{15(x+y)^2}\cdot\dfrac{5x+5y}{x-y}$

$=\dfrac{(x+y)(x-y)}{15(x+y)^2}\cdot\dfrac{5(x+y)}{(x-y)}=\dfrac{1}{3}$

14. $\dfrac{x^2+4x-12}{x^2+9x+18}\div\dfrac{3x+12}{6x+6}=\dfrac{x^2+4x-12}{x^2+9x+18}\cdot\dfrac{6x+6}{3x+12}$

$=\dfrac{(x+6)(x-2)}{(x+6)(x+3)}\cdot\dfrac{6(x+1)}{3(x+4)}=\dfrac{2(x-2)(x+1)}{(x+3)(x+4)}$

15. $\dfrac{z^2+14z+49}{z^2+2z-35}\div\dfrac{z^2+9z+14}{z^2-3z-10}$

$=\dfrac{z^2+14z+49}{z^2+2z-35}\cdot\dfrac{z^2-3z-10}{z^2+9z+14}$

$=\dfrac{(z+7)^2}{(z+7)(z-5)}\cdot\dfrac{(z-5)(z+2)}{(z+7)(z+2)}=1$

16. $\dfrac{x^3y+2x^2y^2+xy^3}{x^4-y^4}\div\dfrac{x^2+xy}{x^2+y^2}$

$=\dfrac{x^3y+2x^2y^2+xy^3}{x^4-y^4}\cdot\dfrac{x^2+y^2}{x^2+xy}$

$=\dfrac{xy(x^2+2xy+y^2)}{x^4-y^4}\cdot\dfrac{x^2+y^2}{x(x+y)}$

$=\dfrac{xy(x+y)^2}{(x+y)(x-y)(x^2+y^2)}\cdot\dfrac{(x^2+y^2)}{x(x+y)}$

$=\dfrac{y}{x-y}$

17. $\dfrac{a+b}{a^2-ab} \div \dfrac{1}{a^2-b^2} = \dfrac{a+b}{a^2-ab} \cdot \dfrac{a^2-b^2}{1}$

$= \dfrac{(a+b)}{a(a-b)} \cdot \dfrac{(a+b)(a-b)}{1} = \dfrac{(a+b)^2}{a}$

18. $\dfrac{x^2+x-20}{4x^3} \div \dfrac{x^2-16}{6x^2} = \dfrac{x^2+x-20}{4x^3} \cdot \dfrac{6x^2}{x^2-16}$

$= \dfrac{(x+5)(x-4)}{4x^3} \cdot \dfrac{6x^2}{(x+4)(x-4)} = \dfrac{3(x+5)}{2x(x+4)}$

19. $\dfrac{7m^2n^2}{8} \div \dfrac{21mn^4}{16} = \dfrac{7m^2n^2}{8} \cdot \dfrac{16}{21mn^4} = \dfrac{2m}{3n^2}$

20. $\dfrac{x^2+6x+9}{x^2+2x-3} \div \dfrac{x^2-9}{x^2-x-6}$

$= \dfrac{x^2+6x+9}{x^2+2x-3} \cdot \dfrac{x^2-x-6}{x^2-9}$

$= \dfrac{(x+3)^2}{(x+3)(x-1)} \cdot \dfrac{(x+2)(x-3)}{(x+3)(x-3)} = \dfrac{x+2}{x-1}$

CHAPTER 18

Adding and Subtracting Rational Expressions

EXERCISE 18.1

1. $\dfrac{5x}{9} + \dfrac{2x}{9} = \dfrac{7x}{9}$

2. $\dfrac{2a}{5y} + \dfrac{3a}{5y} = \dfrac{5a}{5y} = \dfrac{a}{y}$

3. $\dfrac{x}{x+5} + \dfrac{4x}{x+5} = \dfrac{5x}{x+5}$

4. $\dfrac{x}{(x-2)(x+2)} + \dfrac{2}{(x-2)(x+2)}$

$= \dfrac{(x+2)}{(x-2)(x+2)} = \dfrac{1}{(x-2)}$

5. $\dfrac{x^2}{x^2-25} + \dfrac{10x+25}{x^2-25} = \dfrac{x^2+10x+25}{x^2-25}$

$= \dfrac{(x+5)^2}{(x+5)(x-5)} = \dfrac{x+5}{x-5}$

6. $\dfrac{x}{3x+1} + \dfrac{2x}{3x+1} + \dfrac{1}{3x+1} = \dfrac{3x+1}{3x+1} = 1$

7. $\dfrac{2x+2}{5} + \dfrac{2x+1}{5} + \dfrac{x+3}{5} = \dfrac{5x+6}{5}$

8. $\dfrac{x^2+x-1}{x^2+2x+1} + \dfrac{x^2+x+1}{x^2+2x+1} = \dfrac{2x^2+2x}{x^2+2x+1}$

$= \dfrac{2x(x+1)}{(x+1)^2} = \dfrac{2x}{(x+1)}$

9. $\dfrac{5x}{10a} + \dfrac{x}{10a} + \dfrac{4x}{10a} = \dfrac{10x}{10a} = \dfrac{x}{a}$

10. $\dfrac{x^2+5x-13}{x^2+3x-10} + \dfrac{x^2+x-7}{x^2+3x-10}$

$= \dfrac{2x^2+6x-20}{x^2+3x-10} = \dfrac{2(x^2+3x-10)}{x^2+3x-10} = 2$

11. $\dfrac{x}{x^2-16} - \dfrac{4}{x^2-16} = \dfrac{x-4}{x^2-16}$

$= \dfrac{(x-4)}{(x+4)(x-4)} = \dfrac{1}{(x+4)}$

12. $\dfrac{3x}{(x-3)^2} - \dfrac{2x}{(x-3)^2} = \dfrac{3x-2x}{(x-3)^2} = \dfrac{x}{(x-3)^2}$

13. $\dfrac{x^2-xy}{x^2-y^2} - \dfrac{xy-y^2}{x^2-y^2} = \dfrac{x^2-xy-xy+y^2}{x^2-y^2}$

$= \dfrac{x^2-2xy+y^2}{x^2-y^2} = \dfrac{(x-y)^2}{(x+y)(x-y)}$

$= \dfrac{(x-y)}{(x+y)}$

14. $\frac{x^2}{x^2+2x-15}-\frac{25}{x^2+2x-15}=\frac{x^2-25}{x^2+2x-15}$
$=\frac{(x+5)(x-5)}{(x+5)(x-3)}=\frac{(x-5)}{(x-3)}$

15. $\frac{z^2}{z^2-9}-\frac{6z-9}{z^2-9}=\frac{z^2-6z+9}{z^2-9}$
$=\frac{(z-3)^2}{(z+3)(z-3)}=\frac{(z-3)}{(z+3)}$

16. $\frac{7x+4y}{10}-\frac{2x+3y}{10}=\frac{5x+y}{10}$

17. $\frac{x(x-y)}{x^2-y^2}-\frac{y(y-x)}{x^2-y^2}=\frac{x(x-y)-y(y-x)}{x^2-y^2}$
$=\frac{x^2-xy-y^2+xy}{x^2-y^2}=\frac{x^2-y^2}{x^2-y^2}=1$

18. $\frac{4x^2}{x^2-8x+16}-\frac{64}{x^2-8x+16}=\frac{4x^2-64}{x^2-8x+16}$
$=\frac{4(x^2-16)}{x^2-8x+16}=\frac{4(x+4)(x-4)}{(x-4)^2}=\frac{4(x+4)}{(x-4)}$

19. $\frac{7x}{5(x+3)}+\frac{4x}{5(x+3)}-\frac{x}{5(x+3)}$
$=\frac{10x}{5(x+3)}=\frac{2x}{(x+3)}$

20. $\frac{x^2+6x+2}{(x+3)(x-2)}-\frac{2x-1}{(x+3)(x-2)}$
$=\frac{x^2+6x+2-2x+1}{(x+3)(x-2)}$
$=\frac{x^2+4x+3}{(x+3)(x-2)}=\frac{(x+3)(x+1)}{(x+3)(x-2)}$
$=\frac{(x+1)}{(x-2)}$

EXERCISE 18.2

1. $\frac{2x}{5}+\frac{3x}{10}=\frac{7x}{10}$

2. $\frac{5m}{12x^2y}-\frac{3n}{10xy^2}=\frac{25my-18nx}{60x^2y^2}$

3. $\frac{3a+6}{2a}+\frac{5b+4}{2b}=\frac{2a+4ab+3b}{ab}$

4. $\frac{x+y}{2x}-\frac{x-y}{3y}+\frac{y-z}{z}$
$=\frac{-2x^2z-xyz+6xy^2+3y^2z}{6xyz}$

5. $\frac{4(x+2)}{x}-\frac{2(x+4)}{3x}+\frac{x+1}{6x}=\frac{7x+11}{2x}$

6. $\frac{4x^2-5}{3x}+2x=\frac{10x^2-5}{3x}$

7. $4x^2-\frac{12x^2-3}{5x}=\frac{20x^3-12x^2+3}{5x}$

8. $\frac{y(x-3)}{4x}+\frac{x^2-y^2}{xy}-\frac{x(y-2)}{6y}$
$=\frac{-2x^2y+16x^2+3xy^2-21y^2}{12xy}$

9. $\frac{x-4}{4}-\frac{x-4}{x}=\frac{x^2-8x+16}{4x}$

10. $\frac{4}{x+4}+\frac{5}{x-2}=\frac{9x+12}{(x+4)(x-2)}$

11. $\frac{3}{x+1}-\frac{5}{x-2}=\frac{-2x-11}{(x+1)(x-2)}$

12. $\frac{x}{x-y}+\frac{x+y}{xy}=\frac{x^2+x^2y-y^2}{xy(x-y)}$

13. $\frac{3}{x+3}+\frac{2}{x+2}+\frac{2}{3}=\frac{5x+12}{(x+3)(x+2)}$

14. $2x-\frac{3x^2+5}{2x+5}=\frac{x^2+10x-5}{2x+5}$

15. $a+\frac{2a^2}{a-5}-6=\frac{3a^2-11a+30}{a-5}$

16. $\frac{2t}{t+4}-\frac{3t-5}{t^2+8t+16}=\frac{2t^2+5t+5}{(t+4)^2}$

17. $\frac{6z}{z^2+5z+6}-\frac{2z}{z^2+6z+9}=\frac{4z^2+14z}{(z+2)(z+3)^2}$

18. $\dfrac{2x}{x^2-25}+\dfrac{4}{x+5}=\dfrac{6x-20}{(x+5)(x-5)}$

19. $\dfrac{3x-1}{2+4x}-\dfrac{x}{4+2x}=\dfrac{x^2+4x-2}{2(x+2)(2x+1)}$

20. $\dfrac{x-2}{2(x^2-9)}+\dfrac{x+3}{3(x^2-x-6)}$
$=\dfrac{5x^2+12x+6}{6(x+3)(x-3)(x+2)}$

CHAPTER 19

Simplifying Complex Fractions

EXERCISE 19.1

1. $\dfrac{\left(\dfrac{3}{2}\right)}{\left(\dfrac{15}{16}\right)}=\dfrac{8}{5}$

2. $\dfrac{\dfrac{(x^2-y^2)}{8}}{\dfrac{(x-y)}{32}}=4(x+y)$

3. $\dfrac{\left(\dfrac{x}{y^2}\right)}{\left(\dfrac{x^2}{y}\right)}=\dfrac{1}{xy}$

4. $\dfrac{1}{1-\dfrac{x}{y}}=\dfrac{y}{y-x}$

5. $\dfrac{1}{\left(\dfrac{1-x}{y}\right)}=\dfrac{y}{1-x}$

6. $\dfrac{a}{a-\dfrac{a}{2}}=2$

7. $\dfrac{5x^{-1}-\dfrac{3}{4}}{x^{-2}-\dfrac{1}{2}}=\dfrac{20x-3x^2}{4-2x^2}$

8. $\dfrac{x^{-1}+y^{-3}}{x^{-4}+y^{-2}}=\dfrac{x^3y^3+x^4}{y^3+x^4y}$

9. $\dfrac{4x^{-1}y}{\left(\dfrac{z^{-2}}{2}\right)}=\dfrac{8yz^2}{x}$

10. $\dfrac{\left(\dfrac{9a^2b^4}{10c}\right)}{27abc}=\dfrac{ab^3}{30c^2}$

EXERCISE 19.2

1. $\dfrac{\left(\dfrac{4x-4y}{5x+5y}\right)}{\left(\dfrac{4}{25}\right)}=\dfrac{5(x-y)}{x+y}$

2. $\dfrac{\left(\dfrac{4x-4}{x^2-16}\right)}{\left(\dfrac{x-1}{x-4}\right)}=\dfrac{4}{x+4}$

3. $\dfrac{\left(\dfrac{x^2-y^2}{15(x+y)^2}\right)}{\left(\dfrac{x-y}{5x+5y}\right)}=\dfrac{1}{3}$

4. $\dfrac{5-\dfrac{3}{x}}{\dfrac{1}{2}-\dfrac{1}{x^2}}=\dfrac{10x^2-6x}{x^2-2}$

5. $\dfrac{\dfrac{x}{x-3}-\dfrac{x+4}{x+3}}{\dfrac{x}{x^2-9}+\dfrac{6}{x^2-9}}=2$

6. $\dfrac{\dfrac{1}{2x}-\dfrac{1}{4}}{\dfrac{3}{x}+\dfrac{1}{6}}=\dfrac{6-3x}{36+2x}$

7. $\dfrac{\left(\dfrac{a+b}{a^2-ab}\right)}{\left(\dfrac{1}{a^2-b^2}\right)}=\dfrac{(a+b)^2}{a}$

8. $\dfrac{\left(\dfrac{x^2+x-20}{4x^3}\right)}{\left(\dfrac{x^2-16}{6x^2}\right)}=\dfrac{3(x+5)}{2x(x+4)}$

9. $\dfrac{x^{-3}}{y^{-3}-x^{-3}}=\dfrac{y^3}{x^3-y^3}$

10. $\dfrac{\dfrac{x+2}{x-1}-\dfrac{x-3}{x}}{\dfrac{x+4}{x}+\dfrac{x-2}{x-1}}=\dfrac{6x-3}{2x^2+x-4}$

CHAPTER 20

One-Variable Linear Equations and Inequalities

EXERCISE 20.1

1. True
2. False
3. False
4. False
5. True
6. True
7. True
8. True
9. True
10. False
11. False
12. True

13. True
14. False
15. True
16. True
17. False
18. True
19. True
20. True

EXERCISE 20.2

1. $x = \frac{1}{2}$
2. $x = 15$
3. $x = \frac{8}{3}$
4. $x = 11.04$
5. $x = -5$
6. $x = 1$
7. $x = \frac{11}{3}$
8. $x = 2$
9. $x = 5$
10. $x = -3$
11. $x = 24$
12. $x = 48$
13. $x = 8$
14. $x = 9$
15. $x = 840$
16. $x = 84$
17. $x = 4.7$
18. $x = 20$
19. $\frac{x}{2} + \frac{x}{3} = 5,\ x = 6$
20. $\frac{5 + x}{11 + x} = \frac{3}{5},\ x = 4$

EXERCISE 20.3

1. $x = 16y$
2. $y = \frac{3x}{2}$
3. $y = \frac{3}{4}x - \frac{1}{2}$
4. $y = x + 3$
5. $x = y$
6. $r = \frac{C}{2\pi}$
7. $h = \frac{V}{B}$
8. $C = \frac{5}{9}(F - 32)$
9. $W = \frac{P - 2L}{2}$
10. $y = -\frac{1}{2}x + \frac{3}{2}$

EXERCISE 20.4

1. $x > \frac{1}{2}$
2. $x \leq 15$
3. $x > \frac{8}{3}$
4. $x \leq 11.04$
5. $x < -5$
6. $x > 1$
7. $x \geq \frac{11}{3}$
8. $x < 2$
9. $x \geq 5$
10. $x < -3$
11. $x < 24$
12. $x \leq 48$
13. $x \geq 8$
14. $x > 9$
15. $x < 840$

CHAPTER 21

One-Variable Quadratic Equations

EXERCISE 21.1

1. $x^2 - 3x + 2 = 0$
2. $x^2 - 6x + 9 = 0$
3. $x^2 - 8x + 15 = 0$
4. $25x^2 - 10x + 1 = 0$
5. $6x^2 - 17x + 5 = 0$
6. True
7. True
8. False
9. True
10. True
11. True
12. False
13. True
14. True
15. True

EXERCISE 21.2

1. $x = \pm 12$
2. $x = \pm 8$
3. $x = \pm 4\sqrt{2}$
4. $x = \pm\sqrt{7}$
5. $x = \pm 2\sqrt{3}$
6. $x = \pm\frac{1}{2}$
7. $x = \pm\frac{\sqrt{2}}{2}$
8. $x = \pm\frac{3}{4}$

9. $x = \frac{\pm\sqrt{5}}{5}$
10. $x = \pm\sqrt{3}$
11. $x = \pm\sqrt{10}$
12. $x = \pm 6$
13. $x = \pm\sqrt{34}$
14. $x = \pm 5$
15. $x = \pm 2\sqrt{7}$
16. $x = \pm 10$
17. $x = \pm\sqrt{6}$
18. $x = \pm 4$
19. $x = \pm 5$
20. $x = \pm 2\sqrt{7}$

EXERCISE 21.3

1. $x = 3, x = -6$
2. $x = 3, x = -3$
3. $x = 0, x = 7$
4. $x = 0, x = -\frac{5}{4}$
5. $x = 3$
6. $x = 2, x = -9$
7. $x = -\frac{3}{2}, x = -2$
8. $x = 4, x = 1$
9. $x = -4, x = -2$
10. $x = 6, x = -2$
11. $x = 6, x = -3$
12. $x = \frac{9}{7}, x = -\frac{9}{7}$
13. $x = 2, x = -18$
14. $x = -2, x = -4$
15. $x = 3, x = -4$
16. $x = 5, x = -2$
17. $x = 1, x = -2$
18. $x = 5, x = 3$
19. $x = -2, x = -4$
20. $x = 2, x = -12$

EXERCISE 21.4

1. $x = 3, x = -6$
2. $x = 3, x = -3$
3. $x = 7, x = 0$
4. $x = 0, x = -\frac{5}{4}$
5. $x = 3$
6. $x = 2, x = -9$
7. $x = -\frac{3}{2}, x = -2$
8. $x = 4, x = 1$
9. $x = -2, x = -4$
10. $x = 6, x = -2$
11. $x = 4, x = -10$
12. $x = -4 + \sqrt{6}, x = -4 - \sqrt{6}$
13. $x = 2$
14. $x = 2 + \sqrt{14}, x = 2 - \sqrt{14}$
15. $x = 3 + \sqrt{5}, x = 3 - \sqrt{5}$

EXERCISE 21.5

1. $x = 3, x = -6$
2. $x = 3, x = -3$
3. $x = 7, x = 0$
4. $x = 0, x = -\frac{5}{4}$
5. $x = 3$
6. $x = 2, x = -9$
7. $x = -\frac{3}{2}, x = -2$
8. $x = 4, x = 1$
9. $x = -2, x = -4$
10. $x = 6, x = -2$
11. $x = 4, x = -10$
12. $x = -4 + \sqrt{6}, x = -4 - \sqrt{6}$
13. $x = 2$
14. $x = 2 + \sqrt{14}, x = 2 - \sqrt{14}$
15. $x = 3 + \sqrt{5}, x = 3 - \sqrt{5}$
16. $x = 10, x = -10$
17. $x = \sqrt{6}, x = -\sqrt{6}$
18. $x = 4, x = -4$
19. $x = 5, x = -5$
20. $x = 2\sqrt{7}, x = -2\sqrt{7}$

CHAPTER 22

The Cartesian Coordinate Plane

EXERCISE 22.1

1. Real
2. Origin
3. Ordered
4. x-axis, right
5. y-axis, above
6. (−8, 0)
7. (−5, 8)
8. (−3, 3)
9. (0, 4)
10. (8, 6)
11. (−4, −3)
12. (0, −5)
13. (0, 0)
14. (5, −3)
15. (7, 0)

EXERCISE 22.2

1. Counterclockwise
2. Negative
3. Positive
4. y-axis
5. x-axis
6. Quadrant I

7. Quadrant I
8. Quadrant III
9. x-axis
10. Quadrant II
11. Quadrant IV
12. y-axis
13. Quadrant II
14. Quadrant I
15. Quadrant III
16. x-axis
17. y-axis
18. Quadrant IV
19. Quadrant II
20. x-axis

CHAPTER 23

Formulas for the Coordinate Plane

EXERCISE 23.1

1. $\sqrt{53}$
2. $\sqrt{65}$
3. 8
4. $\sqrt{53}$
5. 2
6. $2\sqrt{10}$
7. $\sqrt{53}$
8. $\sqrt{34}$
9. 5
10. $\sqrt{41}$
11. $2\sqrt{5}$
12. $3\sqrt{13}$
13. $2\sqrt{10}$
14. 17
15. 10
16. $\sqrt{10}$
17. 13
18. $\sqrt{14}$
19. $\sqrt{10}$
20. 5

EXERCISE 23.2

1. $\left(5,-\frac{3}{2}\right)$
2. $\left(2,-\frac{1}{2}\right)$
3. $(-2,-3)$
4. $\left(6,\frac{9}{2}\right)$
5. $\left(-\frac{3}{2},-\frac{3}{4}\right)$
6. $(0,0)$

7. $\left(\frac{1}{2}, 6\right)$
8. $\left(\frac{1}{2}, \frac{1}{2}\right)$
9. $\left(\frac{25}{2}, 0\right)$
10. $\left(-3, -\frac{3}{2}\right)$
11. $(-4, -4)$
12. $\left(2, -\frac{3}{2}\right)$
13. $(-2, 1)$
14. $(0.2, 4.8)$
15. $(3, 4)$
16. $\left(-\frac{13}{2}, \frac{1}{2}\right)$
17. $\left(0, \frac{11}{2}\right)$
18. $\left(\frac{\sqrt{7}}{2}, -\frac{\sqrt{7}}{2}\right)$
19. $\left(-1, \frac{9}{2}\right)$
20. $\left(-2, \frac{3}{2}\right)$

EXERCISE 23.3

1. Negative
2. Positive
3. Zero
4. Undefined
5. $-\frac{3}{2}$
6. $-\frac{7}{2}$
7. $-\frac{7}{4}$
8. $\frac{8}{3}$
9. $\frac{7}{2}$
10. 0
11. 3
12. $\frac{2}{7}$
13. $\frac{5}{3}$
14. 0
15. $\frac{5}{4}$
16. $\frac{1}{2}$
17. 7
18. $-\frac{1}{7}$
19. $-\frac{13}{3}$
20. $\frac{3}{13}$

CHAPTER 24

Graphing Lines in the Plane

EXERCISE 24.1

1. (a) -3 (b) 4
2. (a) $\frac{3}{4}$ (b) -1
3. (a) 6 (b) 0
4. (a) $-\frac{4}{3}$ (b) $-\frac{1}{2}$
5. (a) -1 (b) 3
6.

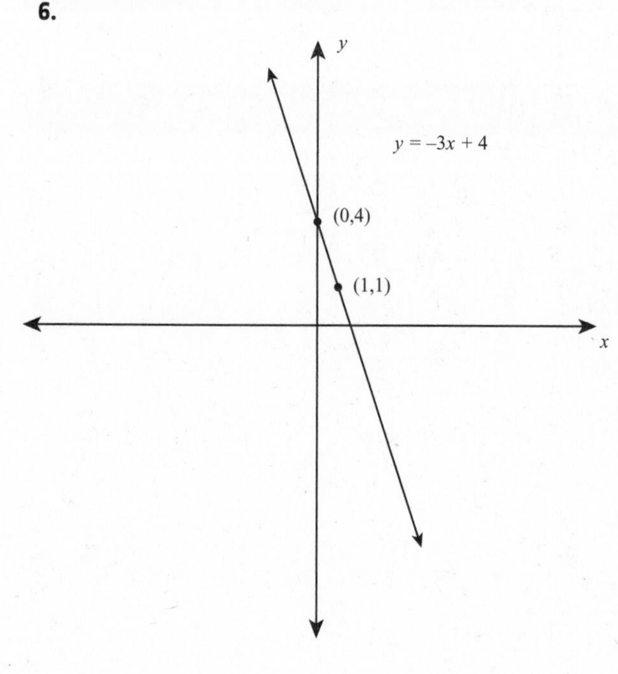

7.

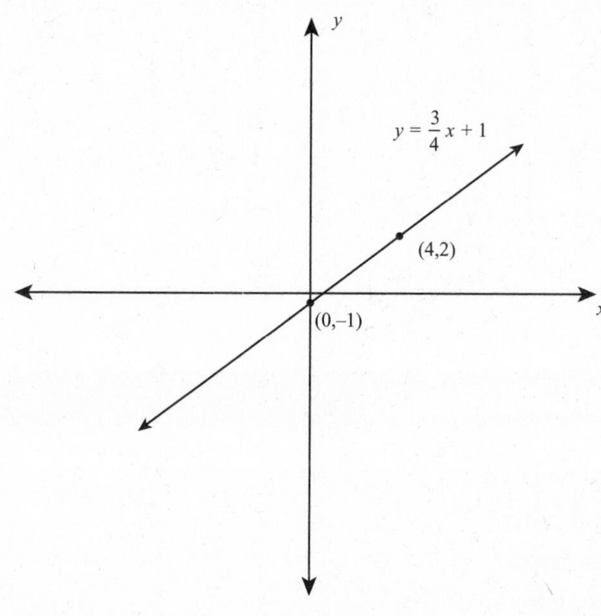

8.

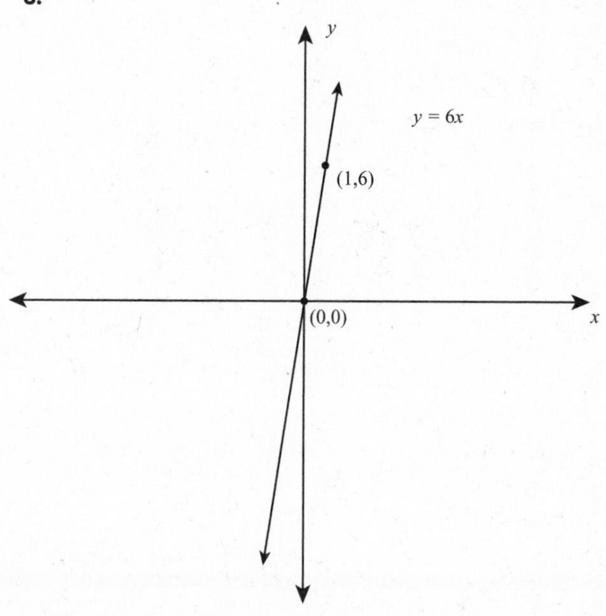

9.

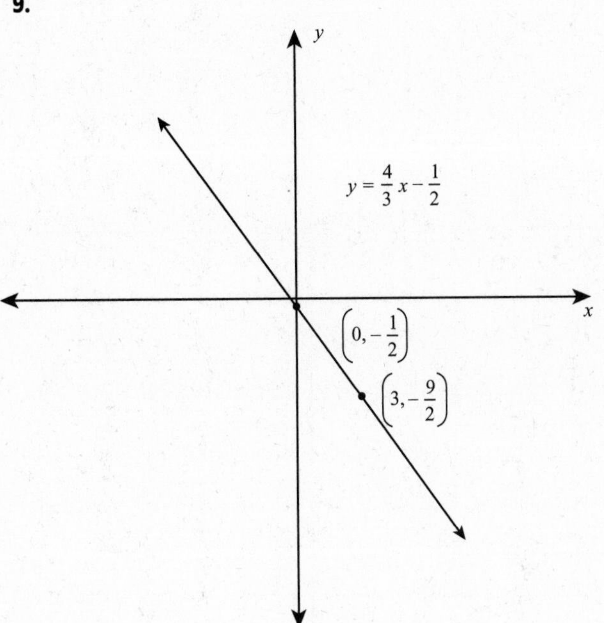
y
x
$y = \frac{4}{3}x - \frac{1}{2}$
$\left(0, -\frac{1}{2}\right)$
$\left(3, -\frac{9}{2}\right)$

10.

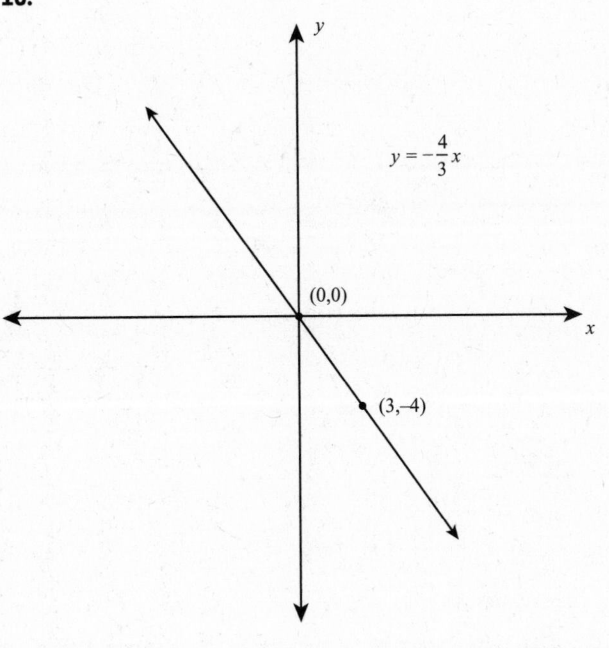
y
x
$y = -\frac{4}{3}x$
(0,0)
(3,–4)

11.

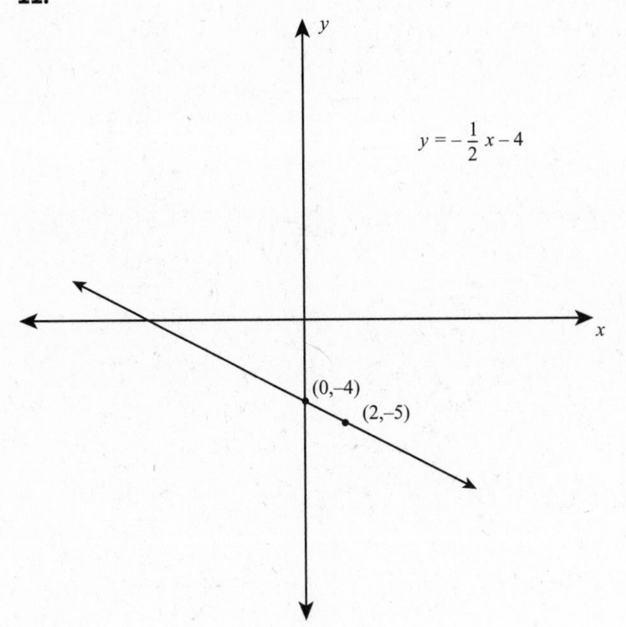
y
x
$y = -\frac{1}{2}x - 4$
(0,–4)
(2,–5)

12.

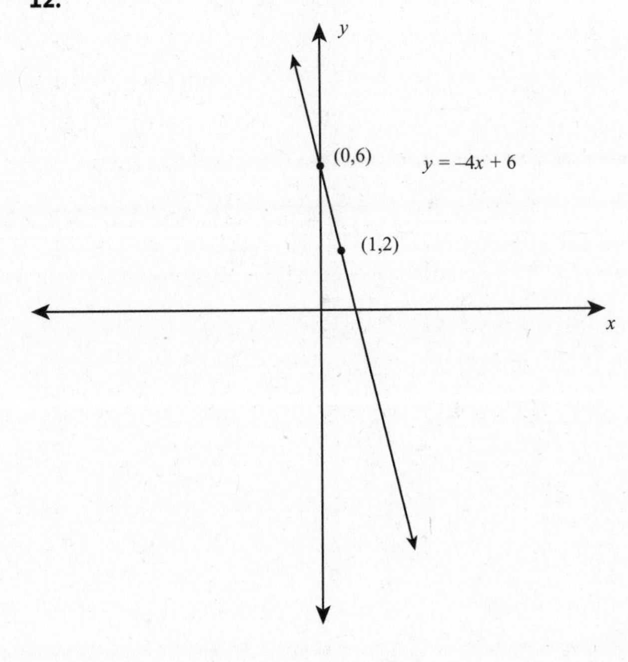
y
x
(0,6)
$y = -4x + 6$
(1,2)

13.

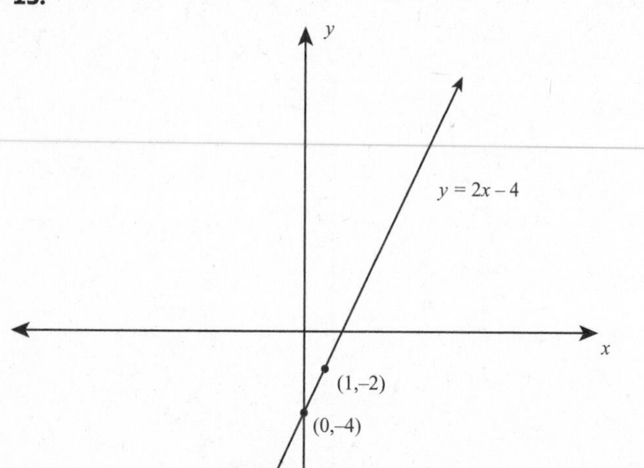

14.

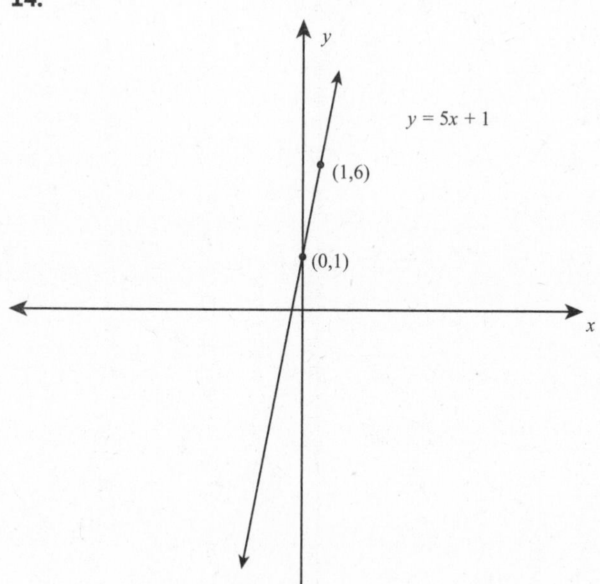

15.

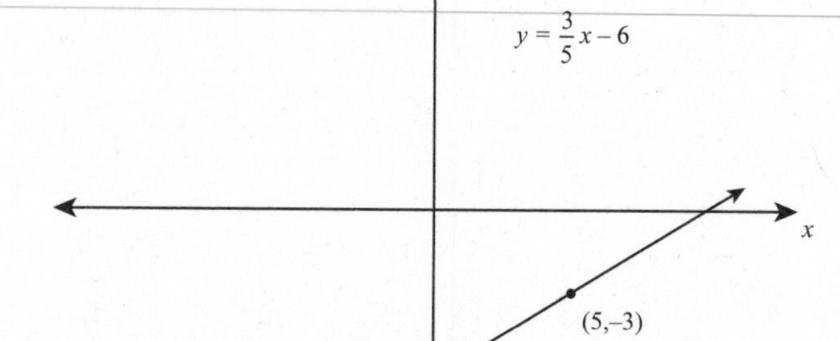

EXERCISE 24.2

1. (a) $\frac{1}{3}$ (b) -3

2. (a) -3 (b) 6

3. (a) -1 (b) 5

4. (a) -5 (b) -4

5. (a) 1 (b) 0

6.

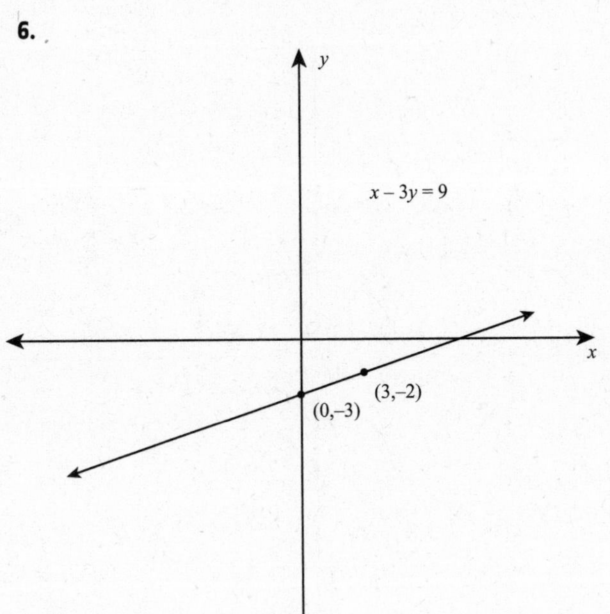

7.

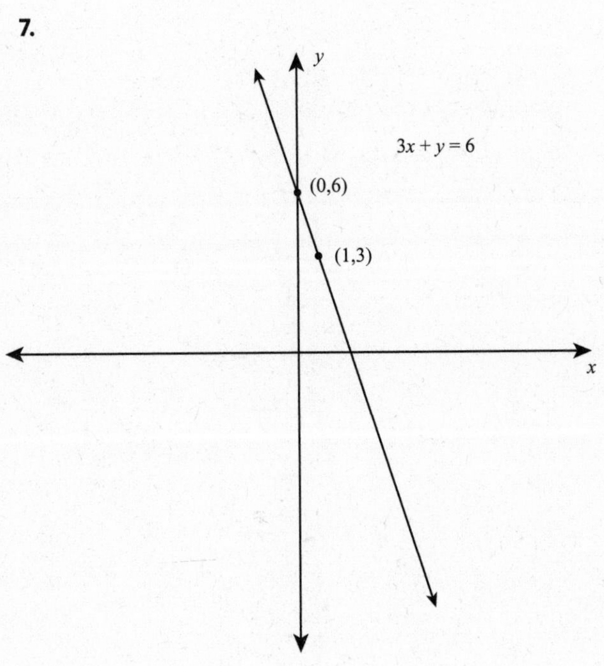

8.

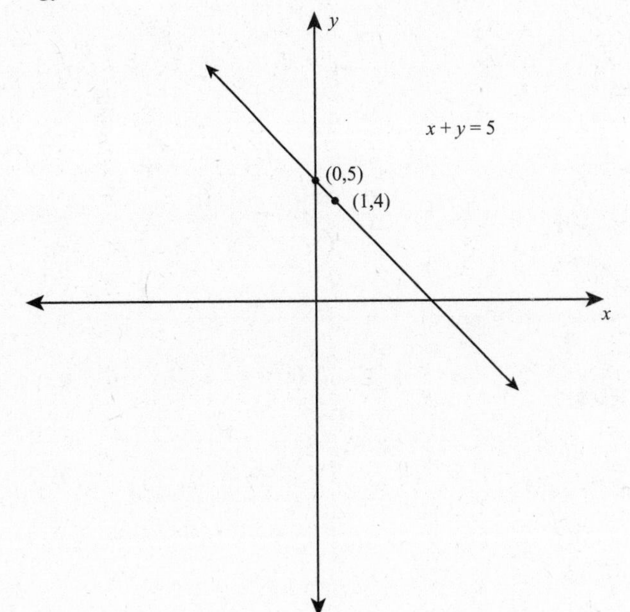

9.

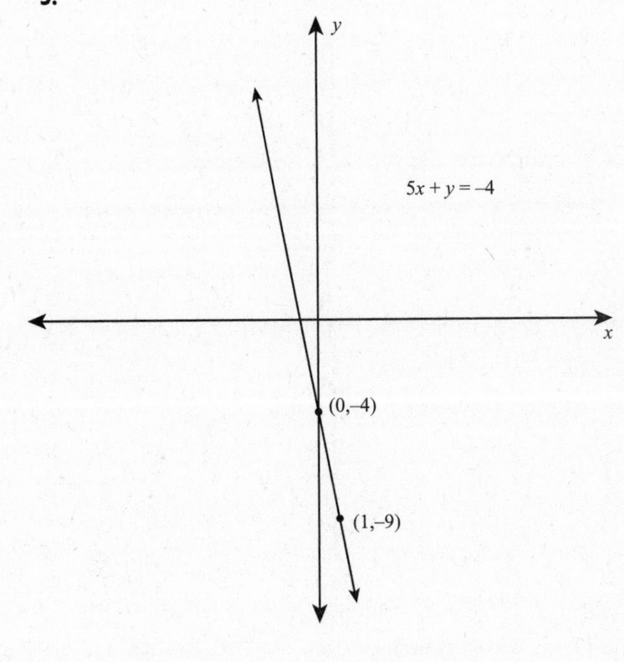

10. **12.**

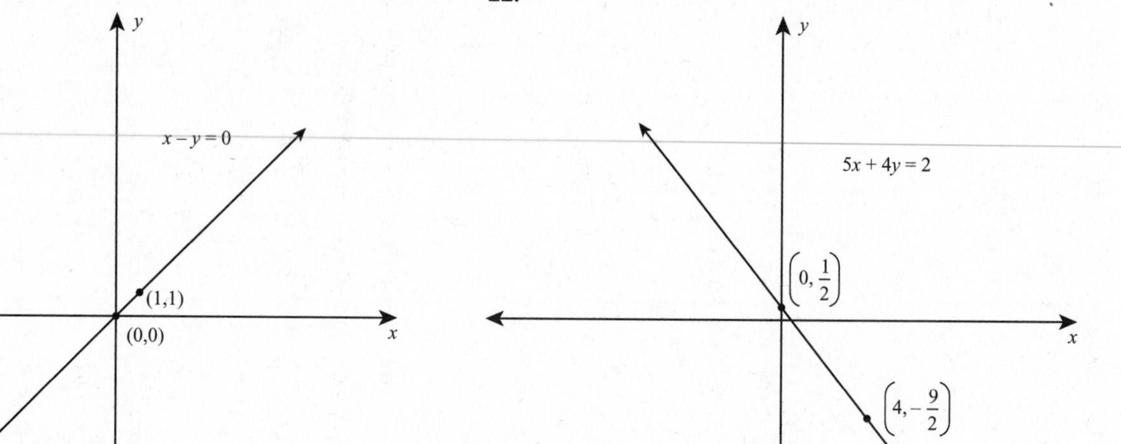

11. **13.**

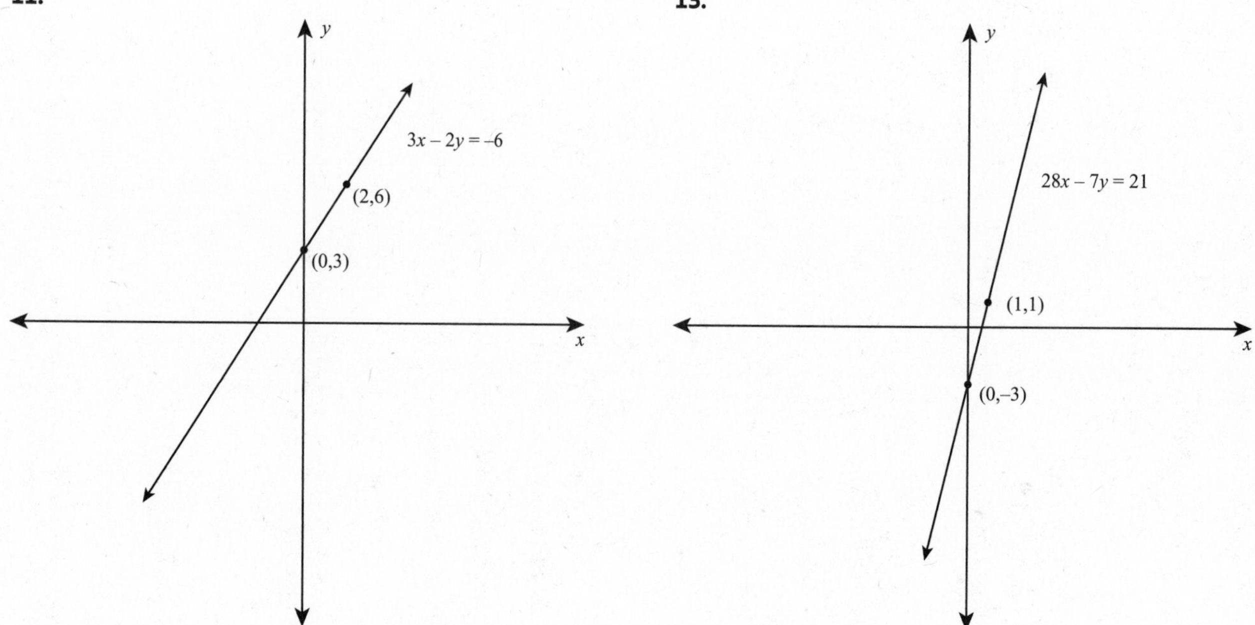

14.

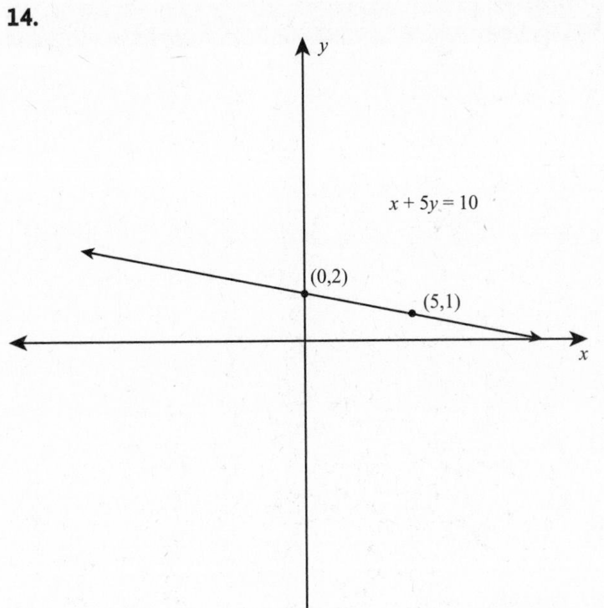

15.

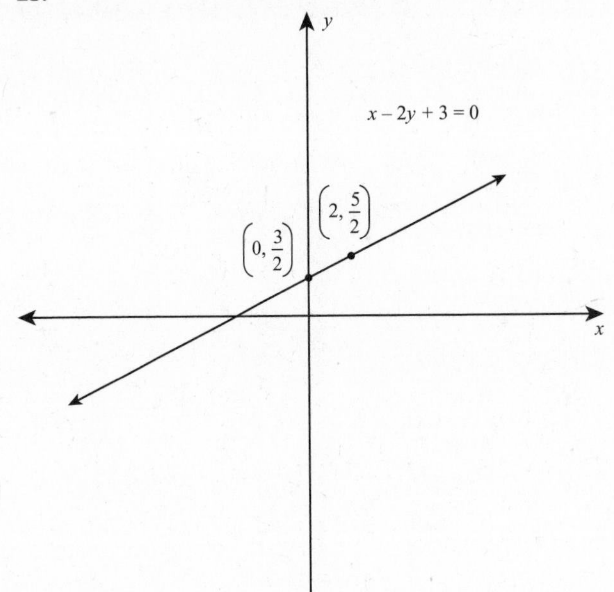

CHAPTER 25

Determining the Equation of a Line

EXERCISE 25.1

1. $y = -x - 5$
2. $y = \frac{1}{2}x + 4$
3. $y = 2x - 5$
4. $y = -\frac{2}{3}x + \frac{5}{3}$
5. $y = 4x$
6. $y = -\frac{3}{8}x - \frac{7}{8}$
7. $y = 5x + 2$
8. $y = -2x + 3$
9. $y = \frac{1}{2}x - 1$
10. $y = -\frac{3}{4}x$
11. $y = -4x + \frac{1}{2}$
12. $y = -\frac{5}{3}x - 3$
13. $y = \frac{7}{4}x - \frac{1}{2}$
14. $y = 4$
15. $y = \frac{3}{8}x - 7$

EXERCISE 25.2

1. $y = 2x - 6$
2. $y = 4x - 2$
3. $y = \frac{1}{2}x - 6$
4. $y = -x - 7$
5. $y = \frac{3}{4}x$
6. $y = -5x + 9$
7. $y = -\frac{3}{4}x + 8$
8. $y = -3x - 24$
9. $y = -\frac{5}{2}x + \frac{1}{2}$
10. $y = \frac{3}{8}x + \frac{47}{8}$
11. $y = \frac{3}{8}x - \frac{9}{4}$
12. $y = -3x - 9$
13. $y = \frac{1}{2}x + \frac{9}{2}$
14. $y = -\frac{4}{3}x - 1$
15. $y = \frac{1}{2}x$
16. $y = 5x + 26$
17. $y = -\frac{5}{3}x - 3$
18. $y = -4x + \frac{1}{2}$
19. $y = \frac{7}{4}x - \frac{1}{2}$
20. $y = 3$

EXERCISE 25.3

1. $y = -\frac{7}{2}x + 16$
2. $y = -\frac{7}{4}x + 3$
3. $y = 4x + 9$
4. $y = \frac{7}{2}x - \frac{33}{2}$
5. $y = -\frac{3}{4}$
6. $y = 3x$
7. $y = \frac{2}{7}x + \frac{41}{7}$
8. $y = \frac{5}{3}x - \frac{1}{3}$
9. $y = 0$
10. $y = \frac{5}{4}x + \frac{9}{4}$
11. $y = \frac{1}{2}x - 2$
12. $y = \frac{3}{2}x - \frac{9}{2}$
13. $y = 3x + 7$
14. $y = 4.8$
15. $y = \frac{4}{3}x$
16. $y = \frac{1}{3}x + \frac{8}{3}$
17. $y = -13x + 12$
18. $y = -x$
19. $y = -3x + \frac{3}{2}$
20. $y = \frac{3}{4}x + 3$

CHAPTER 26

Signal Words and Phrases

EXERCISE 26.1

1. $55x + 200$
2. $5y + 10$
3. $2x + 8y + 9b$
4. $350 + 15x$
5. $125 + 40\%B$
6. $4x + 5$
7. $z + 3z$
8. $m + 15$
9. $3x + 4y$
10. $5x + 20$
11. $9x + 11$
12. $12 + 2x$
13. $a^2 + b^2$
14. $5x + 60$
15. $c + 10\%c$

EXERCISE 26.2

1. $10x - 5y$
2. $80 - 2w$
3. $500 - 20b$
4. $300 - 0.25x$
5. $P - L$
6. $c^2 - a^2$
7. $K - 200$
8. $x - 13$
9. $30 - y$
10. $2x - 10$
11. $5x - 2x$
12. $12 - 7n$
13. $100 - K$
14. $420 - 5y$
15. $6x - 8$

EXERCISE 26.3

1. $5xy$
2. $25 \cdot 3$
3. $8 \cdot 7x$
4. $5\%B$
5. $3y$
6. $2(l + w)$
7. $100b$
8. $\frac{2}{3} \cdot 300x$
9. $0.03x$
10. πd
11. $\frac{1}{2}h(b_1 + b_2)$
12. r^3
13. $50x \cdot 20$
14. $2x^2$
15. y^4

EXERCISE 26.4

1. $\frac{200x}{25y}$
2. $\frac{14p}{7}$
3. $\frac{1{,}500}{10K}$
4. $\frac{(2x-3)}{5}$
5. $\frac{100}{2x}$
6. $\frac{d}{100}$
7. $\frac{400}{0.25x}$
8. $\frac{C}{2\pi r}$
9. $\frac{7x}{2y}$
10. $\frac{a}{b}$
11. $\frac{2x}{100}$
12. $\frac{(5x+6)}{2}$
13. $\frac{(8x)}{\left(\frac{1}{4}\right)}$
14. $\frac{P}{B}$
15. $\frac{600}{t}$

EXERCISE 26.5

1. $2l + 2(l+3) = 52$
2. $(3x-5) = (2x+10)$
3. $25x + 10(300-x) = 4{,}500$
4. $\frac{x}{12} = \frac{200}{3}$
5. $6\%B = 57.60$
6. $0.25q + 0.10(42-q) = 5.55$
7. $55t + 65t = 624$
8. $c^2 = 8^2 + 15^2$
9. $(l+13) = \frac{1}{2}P$
10. $n + 3 = 15$
11. $x + \frac{1}{2}x = 63$
12. $w(w+3) = 70$
13. $95 = 5y + 10$
14. $(n+2) = 2(n+1) + 10$
15. $\frac{1}{3}x + \frac{1}{4}x = 35$

CHAPTER 27

Applying Algebra to Word Problems

EXERCISE 27.1

1. Let n = the number of nickels. Then $759 - n$ = the number of dimes.
2. Let c = the number of pounds of candy that sells at \$11.50 per pound. Then $30 - c$ = the number of pounds of candy that sells at \$19.90 per pound.
3. Let N = Nidhi's age. Then $4N$ = Nidhi's grandmother's age.
4. Let R = Richard's age. Then $R - 5$ = Kat's age.
5. The two numbers are 20 and 52. Hint: Solve $n = 2(72 - n) + 12$.

EXERCISE 27.2

1. $300 \text{ mL} + 200 \text{ mL} = 500 \text{ mL}$
2. $(50 \text{ cm})(20 \text{ cm}) = 1000 \text{ cm}^2$
3. $85° - 10° = 75°$
4. $\frac{4}{3}\pi(6 \text{ in})^3 = \frac{4}{3}\pi(216 \text{ in}^3) = 288\pi \text{ in}^3$
5. $\left(75 \frac{\text{miles}}{\text{hr}}\right)(2 \text{ hr}) = \left(75 \frac{\text{miles}}{\cancel{\text{hr}}}\right)(2 \cancel{\text{hr}})$ $= 300 \text{ miles}$
6. $\frac{13.5 \text{ in}}{0.5 \text{ in}} = \frac{13.5 \cancel{\text{in}}}{0.5 \cancel{\text{in}}} = 26$
7. $\left(\frac{\$25}{\text{hr}}\right)(3.5 \text{ hr}) = \left(\frac{\$25}{\cancel{\text{hr}}}\right)(3.5 \cancel{\text{hr}}) = \$87.50$
8. $2\%(\$1,400) + 1.5\%(\$2,000) = 0.02(\$1,400) + 0.015(\$2,000) = \$28 + \$30 = \$58$
9. $(10 \text{ m})(6 \text{ m})(4 \text{ m}) = 240 \text{ m}^3$
10. $4.5 \text{ lb} \cdot \$15 \text{ per pound} = 4.5 \text{ lb} \cdot \frac{\$15}{\text{lb}} = 4.5 \cancel{\text{lb}} \cdot \frac{\$15}{\cancel{\text{lb}}} = \$67.50$

CHAPTER 28

Applications

EXERCISE 28.1

1. The greatest of the three consecutive integers is -40. Hint: Solve $(n + 2) + 5n = -250$.
2. The three even integers are 20, 22, and 24. Hint: Solve $n + 3(n + 2) = 2(n + 4) + 38$.

3. The number is 24.
   Hint: Solve $2n = (78 - n) - 6$.
4. The number is 4.
   Hint: Solve $x^2 = x + 12$.
5. The number is 21.
   Hint: Solve $n + 8n = 189$.
6. The number is 129.
   Hint: Solve $\frac{2}{3}x = 86$.
7. The number is 112.
   Hint: Solve $x + 0.08x = 120.96$.
8. The number is 260.
   Hint: Solve $x - 0.25x = 195$.
9. The number is 5.
   Hint: Solve $\frac{x}{\left(\frac{1}{2}\right)} = 10$.
10. The number is 100.
   Hint: Solve $\frac{x}{0.25} = 6x - 200$.

EXERCISE 28.2

1. Nidhi is 20 years old.
   Hint: Solve $4N - 10 = 7(N - 10)$.
2. Kat is 15 years old.
   Hint: Solve $(R - 10) = 2[(R - 5) - 10]$.
3. Pablo is 32 years old and his son is 8 years old.
   Hint: Solve $(4s + 16) = 2(s + 16)$.
4. Nathan is 20 years old.
   Hint: Solve $3\left(\frac{1}{5}N + 4\right) = (N + 4)$.
5. Monette is 10 years old, and Juliet is 4 years old.
   Hint: Solve $(J + 6) + 2 = 2(J + 2)$.
6. Samuel is 18 years old.
   Hint: Solve $(S - 3) = 5[(S - 12) - 3]$.
7. Loralei is 12 years old, and Jonah is 7 years old.
   Hint: Solve $8(J - 4) = 3[(J + 5) - 4]$.
8. Liam is 36 years old, and Henri is 9 years old.
   Hint: Solve $(4H - 6) = 10(H - 6)$.
9. Candice is 35 years old, and Sophia is 10 years old.
   Hint: Solve $(45 - S) - 5 = 6(S - 5)$.
10. Arbela is 40 years old, and Loy is 8 years old.
   Hint: Solve $(48 - L) + 8 = 3(L + 8)$.

EXERCISE 28.3

1. The number of women is 24.
   Hint: Solve $4x + 5x = 54$.
2. It takes Raph 20 hours to earn \$485.
   Hint: Solve $\frac{97}{4} = \frac{485}{h}$.
3. It will take 36 weeks for Kenzie to save \$243.
   Hint: Solve $\frac{54}{8} = \frac{243}{w}$.
4. The RV can travel 360 miles.
   Hint: Solve $\frac{270}{18} = \frac{m}{24}$.
5. One partner gets \$1,500, and the other partner gets \$2,000.
   Hint: Solve $3x + 4x = 3{,}500$.
6. The height of the tree is 36 feet.
   Hint: Solve $\frac{h}{30} = \frac{6}{5}$.
7. The number of math teachers attending is 150.
   Hint: Solve $2x + 3x = 375$.
8. The taxes assessed will be \$1,267.
   Hint: Solve $\frac{760}{38{,}000} = \frac{t}{63{,}350}$.

9. It will take 2.5 hours for Baylee to drive 190 miles.

Hint: Solve $\frac{304}{4} = \frac{190}{t}$.

10. The width of the enlarged picture is 10 inches.

Hint: Solve $\frac{4}{w} = \frac{6}{15}$.

EXERCISE 28.4

1. 400 mL of distilled water must be added.

Hint: Solve $50\%(W + 1000) = 70\%(1000)$.

2. The owner should use 10 pounds of the $11.50 candy.

Hint: Solve $11.50C + 19.90(30 - C) = 17.10(30)$.

3. 16 ounces of vinegar must be added.

Hint: Solve $V + 10\%(80) = 25\%(V + 80)$.

4. 700 milliliters of a 10% nitric acid solution must be added.

Hint: Solve $10\%N + 25\%(1400) = 20\%(N + 1400)$.

5. 6 quarts of 100% antifreeze must be added.

Hint: Solve $100\%A + 20\%(10) = 50\%(A + 10)$.

6. 100 quarts of no butterfat milk must be added.

Hint: Solve $4\%(M + 400) = 5\%(400)$.

7. The grocer should use 54 pounds of the $10 nuts and 36 pounds of the $15 nuts.

Hint: Solve $10N + 15(90 - N) = 12(90)$.

8. The manager should use 40 pounds of the cheaper coffee.

Hint: Solve $8C + 14(20) = 10(C + 20)$.

9. 5 liters of a 4% hydrochloric acid solution must be added.

Hint: Solve $4\%H + 20\%(10 - H) = 12\%(10)$.

10. The butcher should use 125 pounds of the 80% lean ground beef and 75 pounds of the 88% lean ground beef.

Hint: Solve $80\%B + 88\%(200 - B) = 83\%(200)$.

EXERCISE 28.5

1. There are 70 dimes in the collection.

Hint: Solve $0.10D + 0.05(200 - D) = 13.50$.

2. There are 16 dimes and 30 nickels.

Hint: Solve $0.10D + 0.05(D + 14) = 3.10$.

3. Ronin has 30 $5 bills and 10 $10 bills.

Hint: Solve $10T + 5(3T) = 250$.

4. Latsha has 16 $5 bills and 4 $1 bills.

Hint: Solve $1D + 5(4D) = 84$.

5. Jermo received 30 dimes.

Hint: Solve $0.25Q + 0.10(Q + 2) = 10$.

6. Barb has 18 dimes, 6 nickels, and 24 pennies.

Hint: Solve $0.10(3N) + 0.05N + 0.01(3N + 6) = 2.34$.

7. Joyce has 3 quarters.

Hint: Solve $0.25(Q) + 0.05(16 - Q) = 1.40$.

8. Willow has 18 nickels.

Hint: Solve $0.05(N) + 0.10(52 - N) = 4.30$.

9. Cael has 24 quarters and 30 nickels.

Hint: Solve $0.25(Q) + 0.05(Q + 6) = 7.50$.

10. Scarlett has 20 quarters.

Hint: Solve $0.25[45 - N - (N + 5)] + 0.10(N + 5) + 0.05N = 7.00$.

EXERCISE 28.6

1. The two vehicles will pass each other in 2 hours.
   Hint: Solve $70t + 65t = 270$.
2. The two vehicles will be 325 miles apart at 10:30 p.m. of the same day.
   Hint: Solve $70t + 60t = 325$.
3. The two vehicles will arrive in the same location in 4.5 hours.
   Hint: Solve $65t + 55t = 540$.
4. The car traveled at an average speed of 80 miles per hour.
   Hint: Solve $3r = (60)(4)$.
5. The two trains will be 315 miles apart in 3 hours.
   Hint: Solve $55t + 50t = 315$.
6. The airplane will take 0.25 hours, which is 15 minutes, to cover a distance of 137.5 miles.
   Hint: Solve $550t = 137.5$.
7. Mora's average speed is 4 miles per hour.
   Hint: Solve $8r + 4r = 24$.
8. The two bicycle riders will meet after 1.5 hours.
   Hint: Solve $16t + 14t = 45$.
9. The boat's speed when there is no current is 21 miles per hour.
   Hint: Solve $3(r + 3) = 4(r - 3)$.
10. The car's average speed is 50 miles per hour.
    Hint: Solve $2(2r - 30) = 2r + 20$.

EXERCISE 28.7

1. It will take 1.875 hours for the two pipes together to fill the tank.
   Hint: Solve $\frac{1}{5} + \frac{1}{3} = \frac{1}{t}$.
2. It takes 16 minutes for the two of them working together to mow the lawn.
   Hint: Solve $\frac{1}{24} + \frac{1}{48} = \frac{1}{t}$.
3. It would take Kelsey 3 hours to mow the lawn alone, and it would take Imogene 6 hours to mow the lawn alone.
   Hint: Solve $\frac{1}{r} + \frac{1}{2r} = \frac{1}{2}$.
4. It will take 24 minutes to water the lawn if both sprinklers operate at the same time.
   Hint: Solve $\frac{1}{40} + \frac{1}{60} = \frac{1}{t}$.
5. It will it take 80 hours to fill the tank if the drain is open while the tank is being filled.
   Hint: Solve $\frac{1}{16} - \frac{1}{20} = \frac{1}{t}$.
6. It will take the two sisters 60 minutes to wash the family car.
   Hint: Solve $\frac{1}{100} + \frac{1}{150} = \frac{1}{t}$.
7. It will take the two machines 0.75 hours to produce 500 items.
   Hint: Solve $\frac{1}{3} + \frac{1}{1} = \frac{1}{t}$.
8. It took a total of 7 hours to complete the job.
   Hint: Solve $\frac{1}{9}t + \frac{1}{18}t = \frac{2}{3}$. Then compute $t + 3$.
9. Madison's roommate paints $\frac{1}{12}$ of the room per hour.
   Hint: Solve $\frac{1}{4} + \frac{1}{r} = \frac{1}{3}$.
10. It will take 24 hours to fill the tank if both pipes are open.
    Hint: Solve $\frac{1}{8} - \frac{1}{12} = \frac{1}{t}$.

EXERCISE 28.8

1. \$44.25 is saved.
   Hint: Solve $P = 15\%(295)$.
2. Ash's total sales last week were \$1,845.
   Hint: Solve $55.35 = 3\%B$.
3. The amount is 28% of the original price.
   Hint: Solve $1{,}624 = R(5{,}800)$.
4. The regular price of the watch was \$288.
   Hint: Solve $w - 25\%w = 216$.
5. The dealer paid \$500 for the television.
   Hint: Solve $c + 30\%c = 650$.
6. The amount of Jaylynn's sales was \$21,250.
   Hint: Solve $(1{,}450 - 600) = 4\%S$.
7. The number is 50.
   Hint: Solve $x + 4\%x = 52$.
8. The original price is \$60.
   Hint: Solve $p + 250\%p = 210$.
9. The sales tax rate is 8%.
   Hint: Solve $7.84 = R(98)$.
10. \$677.66 is 109.3% of \$620.
    Hint: Solve $P = 109.3\%(620)$.

EXERCISE 28.9

1. The interest earned is \$1,800.
   Hint: Solve $I = (15{,}000)(1.5\%)(8)$.
2. It will take 4 years.
   Hint: Solve $400 = (5{,}000)(2\%)(t)$.
3. The investment earns \$288.
   Hint: Solve $I = (4{,}800)(2\%)(3)$.
4. The simple interest rate per year is 1.5%.
   Hint: Solve $262.50 = (3{,}500)(r)(5)$.
5. The amount invested at 4% is \$3,500, and the amount invested at 3% is \$5,800.
   Hint: Solve $4\%x = 3\%(9{,}300 - x) - 34$.
6. The principal is \$3,900.
   Hint: Solve $156 = (P)(2\%)(2)$.
7. The amount invested at 2% is \$1,500, and the amount invested at 4% is \$3,000.
   Hint: Solve $2\%x + 4\%(2x) = 150$.
8. The amount invested at 2% is \$4,200 and the amount invested at 3% is \$2,800.
   Hint: Solve $2\%x = 3\%(7{,}000 - x)$.
9. The certificate of deposit will earn \$62.50 in interest in one year.
   Hint: Solve $P = (2{,}500)(2.5\%)(1)$.
10. Ace will owe \$2,720 to his friend.
    Hint: Solve $x = 2{,}000 + (2{,}000)(12\%)(3)$.

EXERCISE 28.10

1. The fence's length is 18 meters, and its width is 9 meters.
   Hint: Solve $W(2W) = 162$.
2. The flower box's width is 8 inches.
   Hint: Solve $(36)(w)(6) = 1{,}728$.
3. The area of the field is 175,000 feet^2.
   Hint: Solve $2(500) + 2w = 1{,}700$. Then calculate $500w$ to determine the area.
4. The length of the triangle's base is 12 inches.
   Hint: Solve $\frac{1}{2}(b)(18) = 108$.
5. The approximate area is 314 inch^2.
   Hint: Solve $\pi d = 20\pi$. Then calculate $A \approx 3.14\left(\frac{d}{2}\right)^2$.

6. The measure of the third angle is 75°.

Hint: Solve $A + 42 + 63 = 180$.

7. The length of the other leg is 30 centimeters.

Hint: Solve $a^2 + 16^2 = 34^2$.

8. The width of the play area is 5 feet.

Hint: Solve $W(W + 7) = 60$.

9. The diameter is 16 meters.

Hint: Solve $\pi r^2 = 64\pi$. Then calculate $d = 2r$.

10. The rectangle's length is 10 centimeters, and its width is 8 centimeters.

Hint: Solve $2L + 2(L - 2) = 36$.

CHAPTER 29

Introduction to Functions

EXERCISE 29.1

1. Yes

2. No

3. Yes

4. Yes

5. Yes

6. Yes

7. Yes

8. Yes

9. Yes

10. No

11. Domain: $\{-5, -2, 0, 2, 5\}$; Range: $\{-5, -2, 0, 2, 5\}$

12. Domain: $\left\{-\frac{1}{2}, \frac{1}{2}\right\}$; Range: $\{0\}$

13. Domain: $\{-3, -2, -1, 0, 1, 2, 3\}$; Range: $\{-1, 0, 3, 8\}$

14. Domain: $\{-3, -2, -1, 0, 1\}$; Range: $\{-4, -3, 0, 5\}$

15. Domain: $\{-4, 2, 5, 9\}$; Range: $\{-29.5, -11.5, 2, 29\}$

16. Domain: $\{1, 2, 3, 4, \ldots\}$; Range: $\{5, 10, 15, 20, \ldots\}$

17. Domain: $\{-8, -6, 2, 4, 10\}$; Range: $\{-51, -24, -15, -6, 30\}$

18. Domain: $\{1, 2, 3, 4, 5, \ldots\}$; Range: $\{4, 8, 12, 16, 20, \cdots\}$

19. Domain: $\left\{\frac{1}{2}\right\}$; Range: $\{-5\}$

20. Domain: $\{1, 2, 4, 6, 9\}$; Range: $\{1.0, 1.4, 2.2, 3.0, 3.5\}$

EXERCISE 29.2

1. 4

2. 5

3. 5.25

4. 15

5. −3

6. 0

7. 25

8. 50

9. 1,600

10. $\sqrt{3}$

11. $\sqrt{5}$
12. 3
13. -1
14. 5
15. 1
16. R
17. R, $x \neq 3$
18. R
19. R
20. R, $x \geq \frac{1}{2}$

CHAPTER 30

Graphs of Functions

EXERCISE 30.1

1. No
2. Yes
3. Yes
4. Yes
5. Yes
6. No
7. Yes
8. Yes
9. No
10. No

EXERCISE 30.2

1. (a) zero: -5; x-intercept: -5 (b) y-intercept: 10
2. (a) zeros: $-3, -2$; x-intercepts: $-3, -2$ (b) y-intercept: 6
3. (a) zero: none; x-intercept: none (b) y-intercept: 1,500
4. (a) zero: 0; x-intercept: 0 (b) y-intercept: 0
5. (a) zero: 0; x-intercept: 0 (b) y-intercept: 0
6. (a) zero: none; x-intercept: none (b) y-intercept: 3
7. (a) zero: 0; x-intercept: 0 (b) y-intercept: 0
8. (a) zeros: $-\frac{1}{2}$, 3; x-intercepts: $-\frac{1}{2}$, 3 (b) y-intercept: -3
9. (a) zero: none; x-intercept: none (b) y-intercept: $\frac{1}{125}$
10. (a) zero: 32; x-intercept: 32 (b) y-intercept: 8

EXERCISE 30.3

1. Decreasing $x < -3$; Increasing $x > -3$
2. Increasing $x < 2$; Decreasing $x > 2$
3. Decreasing on R
4. Constant on R

5. Increasing $x < -3$; Decreasing x between -3 and 1; Increasing $x > 1$
6. Increasing $x < -3$; Decreasing $x > -3$
7. Increasing on R
8. Increasing $x < -4$; Decreasing x between -4 and 2; Increasing $x > 2$
9. Decreasing on R
10. Increasing $x < -2$; Constant x between -2 and 3; Decreasing $x > 3$

CHAPTER 31

Common Functions and Their Graphs

EXERCISE 31.1

1. Nonvertical, -4.5, 9, 2, decreasing
2. Nonvertical, 75, 0, 0, increasing
3. Nonvertical, $-\frac{1}{5}$, 0, 0, decreasing
4. Nonvertical, -3, 72, 24, decreasing
5. Nonvertical, $-\frac{4}{3}$, 6, $\frac{9}{2}$, decreasing
6. -3
7. 6
8. \$15 per shirt
9. 65 miles per hour
10. 8-gallon decrease per hour

EXERCISE 31.2

1. Parabola, $\left(\frac{5}{4}, -\frac{49}{8}\right)$, -3, upward, minimum, $x = \frac{5}{4}$
2. Parabola, (0,100), 100, downward, maximum, $x = 0$
3. Parabola, $(-5, 0)$, 25, upward, minimum, $x = -5$
4. Parabola, $(-1, 9)$, 8, downward, maximum, $x = -1$
5. Zeros: $-\frac{1}{2}$, 3; x-intercepts: $-\frac{1}{2}$, 3
6. Zeros: -10, 10; x-intercepts: -10, 10
7. Zero: -5; x-intercept: -5
8. Zeros: -4, 2; x-intercepts: -4, 2
9. Zeros: -3, 3; x-intercepts: -3, 3
10.

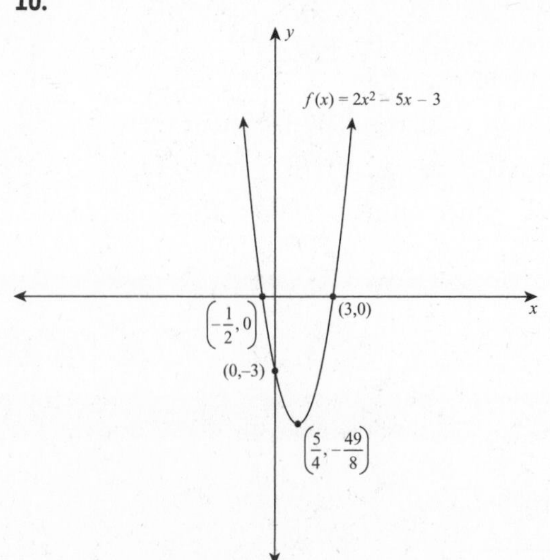

11.

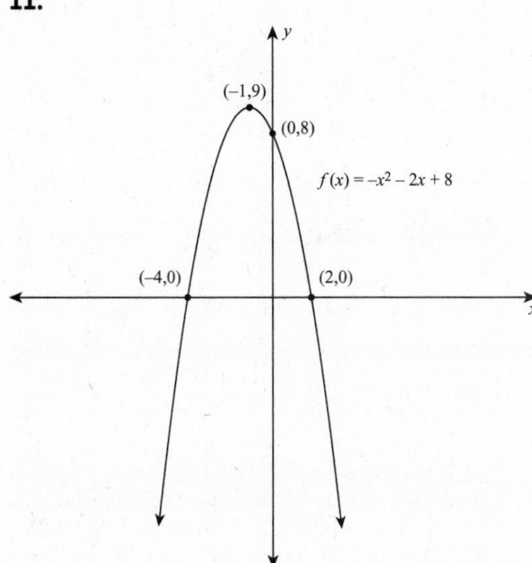

12. 84 feet
    Hint: Evaluate $h(3)$.
13. 2 seconds
    Hint: The vertex is (2, 100).
14. 100 feet
    Hint: The vertex is (2, 100).
15. 4.5 seconds
    Hint: Solve $h(t) = 0$.

EXERCISE 31.3

1. $\frac{1}{25}$
2. 1
3. 25
4. 16
5. 1
6. $\frac{1}{16}$
7. 4
8. 1
9. $\frac{1}{4}$
10. 1,000
11. 2,000
12. 32,000
13. 400 million
14. $1,560.60
15. $5,932.62

CHAPTER 32

Introduction to Systems of Two-Variable Equations

EXERCISE 32.1

1. No, because (–2,1) does not satisfy both equations.
2. Yes, because (2, 5) satisfies both equations simultaneously.
3. Yes, because $\left(\frac{1}{2},3\right)$ satisfies both equations simultaneously.

4. No, because (0, 2) does not satisfy both equations.
5. Yes, because (3, 0) satisfies both equations simultaneously.
6. $\begin{aligned} 8x - 2y &= 6 \\ x - 3y &= -13 \end{aligned}$
7. $\begin{aligned} 2x + y &= 4 \\ 2x - 3y &= -8 \end{aligned}$
8. $\begin{aligned} 4x + 2y &= 8 \\ 2x + y &= -8 \end{aligned}$
9. $\begin{aligned} 3x - 2y &= -3 \\ 6x + 2y &= 9 \end{aligned}$
10. $\begin{aligned} 7x + 14y &= 2 \\ 14x - 7y &= -11 \end{aligned}$

EXERCISE 32.2

1. One solution
2. No solution
3. One solution
4. One solution
5. Infinitely many solutions
6. Intersecting
7. Parallel
8. Intersecting
9. Intersecting
10. Coincident

CHAPTER 33

Solving Systems of Two-Variable Linear Equations

EXERCISE 33.1

1. (2, 3)
2. (2, 5)
3. $\left(\frac{1}{2}, 3\right)$
4. (2, 3)
5. Infinitely many solutions
6. (2, 5)
7. $\left(\frac{1}{2}, 3\right)$
8. No solution
9. $\left(\frac{2}{3}, \frac{5}{2}\right)$
10. $\left(-\frac{4}{7}, \frac{3}{7}\right)$
11. Infinitely many solutions
12. (1, –1)
13. No solution
14. (3, 2)
15. (13.5, 11.5)

Let x = the greater number, and y = the lesser number. Solve the system.

$$\begin{aligned} x + y &= 25 \\ x - y &= 2 \end{aligned}$$

$x = 13.5, y = 11.5$

EXERCISE 33.2

1. (2, 3)
2. (2, 5)
3. $\left(\frac{1}{2}, 3\right)$
4. (2, 3)
5. Infinitely many solutions
6. (2, 5)
7. $\left(\frac{1}{2}, 3\right)$
8. No solution
9. $\left(\frac{2}{3}, \frac{5}{2}\right)$
10. $\left(-\frac{4}{7}, \frac{3}{7}\right)$
11. Infinitely many solutions
12. (1, −1)
13. No solution
14. (3, 2)
15. 19 sheep, 23 chickens

Let S = the number of sheep, and C = the number of chickens. Solve the system.

$$2S + 2C = 84$$
$$4S + 2C = 122$$

$S = 19$, $C = 23$

EXERCISE 33.3

1. (2, 3)
2. (2, 5)
3. $\left(\frac{1}{2}, 3\right)$
4. (2, 3)
5. (2, 5)
6. $\left(\frac{1}{2}, 3\right)$
7. $\left(\frac{7}{8}, \frac{3}{4}\right)$
8. (1, −1)
9. $\left(\frac{2}{3}, \frac{5}{2}\right)$
10. No graph is necessary. A quick check of the coefficients reveals that the lines representing the equations are parallel, so there is no solution.

CHAPTER 34

Graphing Systems of Two-Variable Inequalities

EXERCISE 34.1

1. Above
2. Below
3. Below
4. Above
5. Below

6. $x + y > -5$

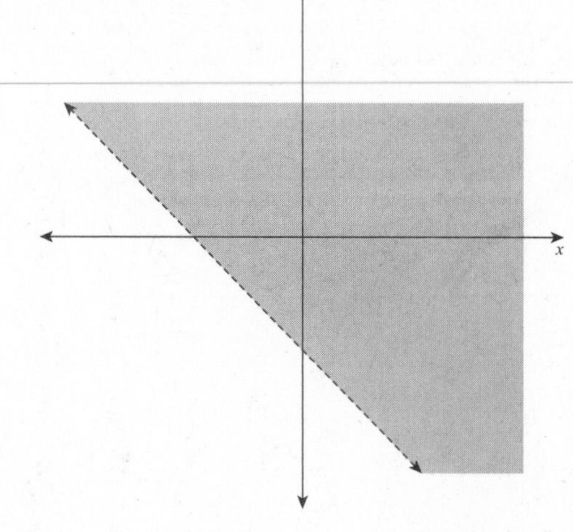

7. $\frac{3}{2}x - y > 5$

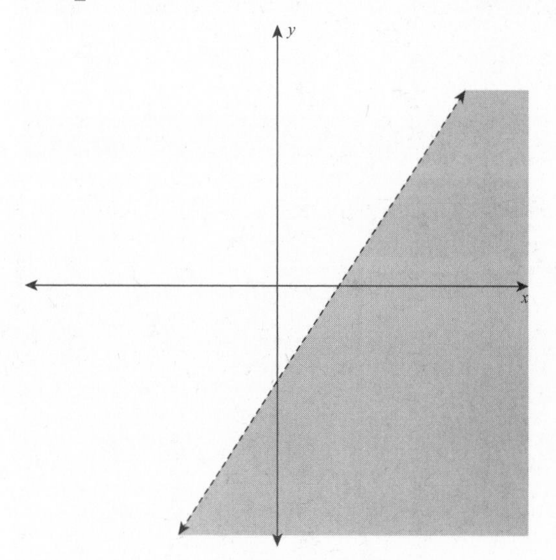

8. $x - y \geq 4$

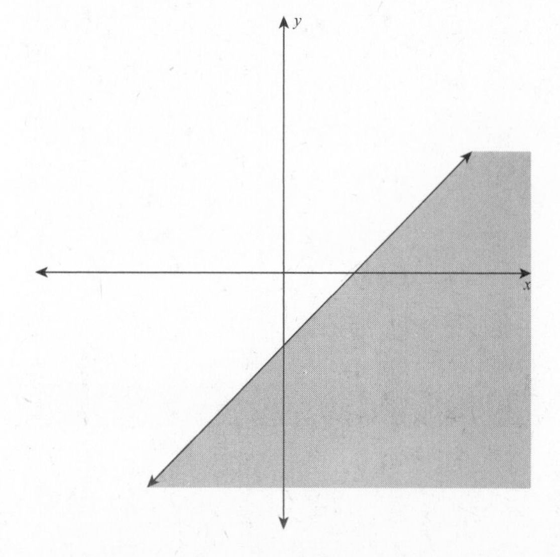

9. $4x - 3y \leq 12$

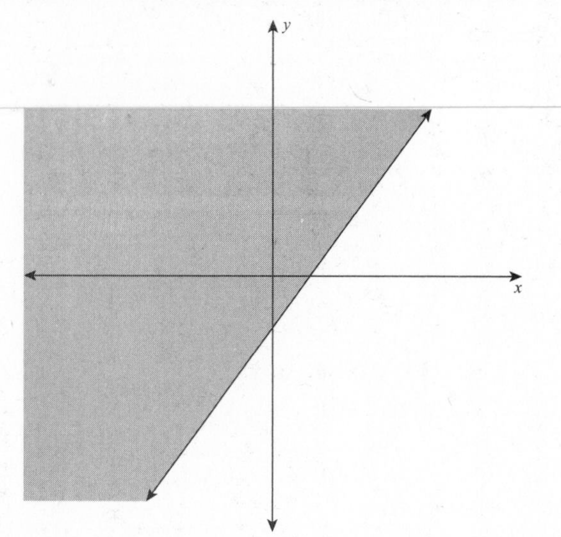

10. $4 - y > 0$

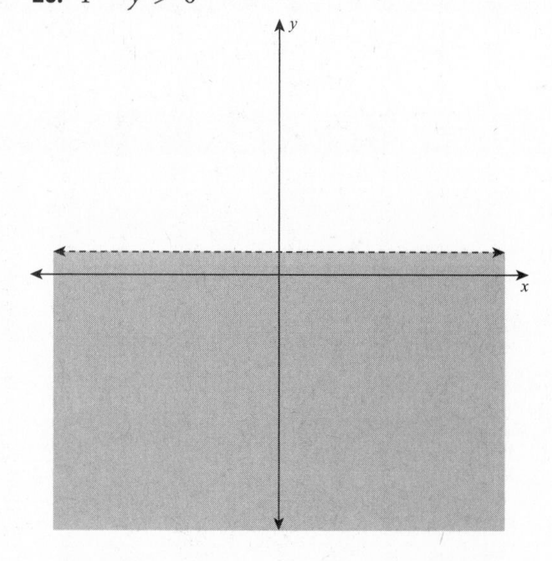

EXERCISE 34.2

1. Above
2. Below
3. Below
4. Above
5. Above
6. $y > 4x^2$

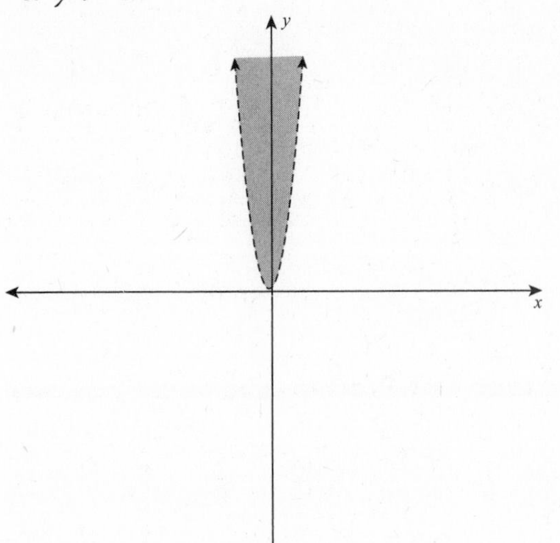

7. $2x^2 + y \leq 8$

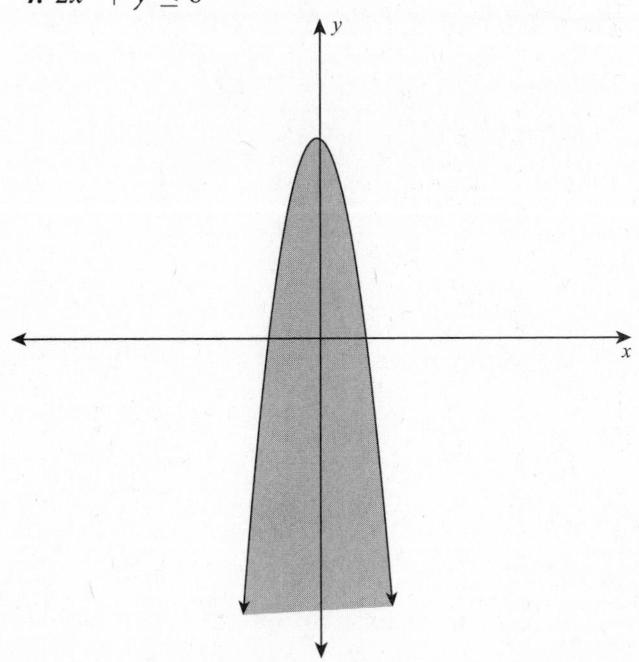

8. $x^2 > y + 1$

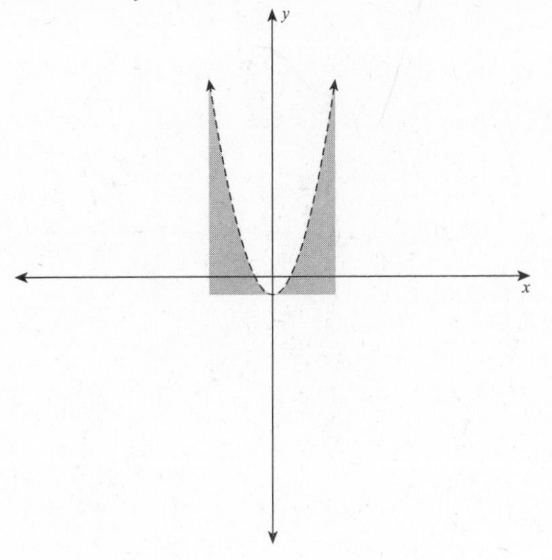

9. $2x^2 - 5x < y + 3$

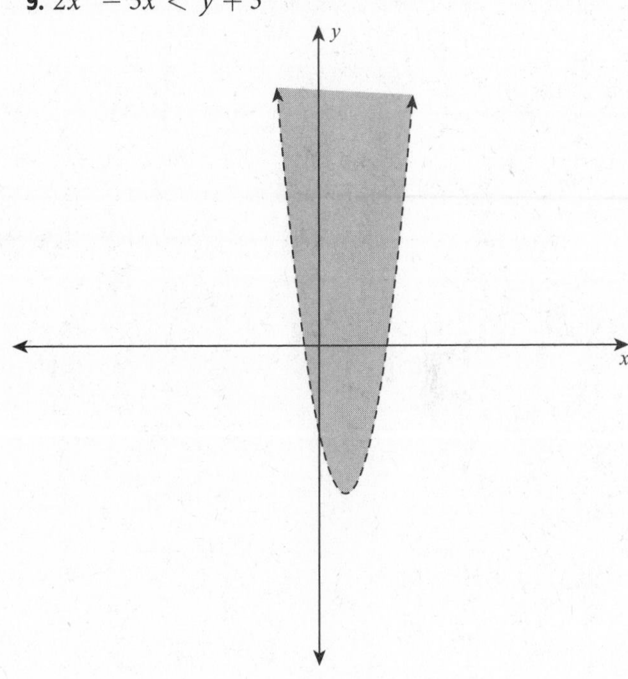

10. $x^2 - y < 9$

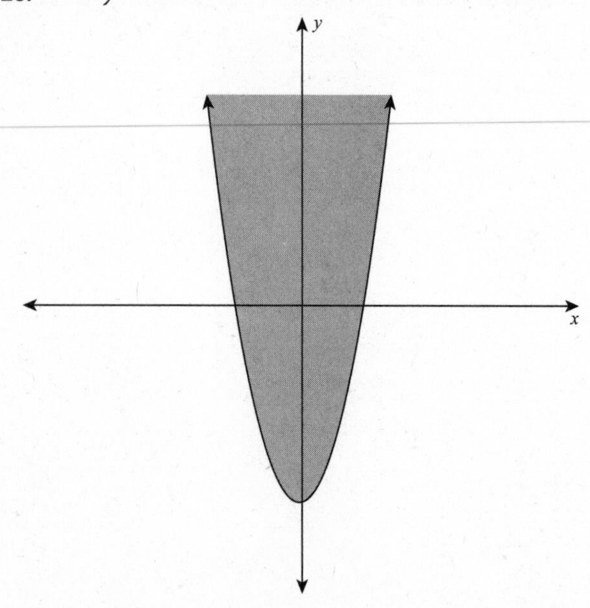